NATURAL NOVELTY

The Newness Manifest in Existence

Richard Boyle

University Press of America,® Inc.
Lanham · Boulder · New York · Toronto · Plymouth, UK

University Press of America,® Inc.
4501 Forbes Boulevard
Suite 200
Lanham, Maryland 20706
UPA Acquisitions Department (301) 459-3366

Unit A, Whitacre Mews, 26-34 Stannary Street,
London SE11 4AB, United Kingdom

Printed in the United States of America
British Library Cataloging in Publication Information Available

Library of Congress Control Number: 2015955032
ISBN: 978-0-7618-6708-1 (clothbound : alk. paper)
eISBN: 978-0-7618-6709-8

Cover photo courtesy of NASA

∞™ The paper used in this publication meets the minimum
requirements of American National Standard for Information
Sciences—Permanence of Paper for Printed Library Materials,
ANSI Z39.48-1992

"There was neither non-existence nor existence then. There was neither the realm of space nor the sky which is beyond. What stirred? Where?"
Creation hymn of the Rigveda, ~1700B.C.[1]

"There is one story left, one road: that it is. And on this road there are very many signs that being is uncreated and imperishable, whole, unique, unwavering, and complete."
Parmenides ~5th century B.C.[2]

"What has been is what will be, and what has been done is what will be done, and there is nothing new under the sun. Is there a thing of which it is said, 'See, this is new?' It has been already in the ages before us."
Ecclesiastes 1 : 4 – 11, Old Testament, ~ 3rd Century B.C[3].

"There is no possible way of making an inference from the existence of one situation to the existence of another, entirely different situation. There is no causal nexus to justify such an inference. We cannot infer the events of the future from those of the present. Superstition is nothing but belief in the causal nexus."
Ludwig Wittgenstein,1918[4]

"Why are there beings at all and why not rather nothing? That is the question."
Martin Heidegger,1929[5]

CONTENTS

PREFACE	1
ACKNOWLEDGMENTS	4
CHAPTER SUMMARIES	5
PRELUDE: RECURRENCE AND RECOMBINATION	12
A FORMAL ABSTRACT SCHEME FOR NOVELTY	27
Summary in words	28
Expansion using a logical formalism	32
THEORETICAL CONSIDERATIONS RELATING TO REAL INSTANCES OF NOVELTY DURING THE HISTORY OF LIFE ON EARTH	67
Life: Nature's most novel process	68
Information and homeostasis	75
The origin of heritable variation in fitness	77
Consciousness: The most novel property of life	103
Earth: The most novel known planet	110
Conclusion	129
NOVELTY, MEANINGFUL LANGUAGE, SUBJECTIVITY AND TIME: A COMPARATIVE READING OF SELECTED WORKS BY LUDWIG WITTGENSTEIN AND MARTIN HEIDEGGER	134

Wittgenstein 142

***Tractatus Logico Philosophicus*: Kicking away a ladder of nonsense** 144

***Philosophical investigations*: Meaning is use, use is contextual and evolving** 156

***Remarks on the foundations of mathematics*: Foundationless useful conventions** 162

***Zettel*: Mind, body and language** 170

***On Certainty*: The relativity of knowledge to the possibility of doubt** 175

Heidegger 182

Being and Time 184

Previous comparison between Wittgenstein and Heidegger: *Groundless Grounds* by Lee Braver 196

Reflections and suggestions 200

Novelty necessitates the evolution of meaningful language over time 205

Novelty is the reason that we can sometimes think what we cannot say 212

The boundary separating the describable from the ineffable changes over time, due to the reality of novelty 227

Novelty explains why there is an aspect of disclosure to truth 236

CONCLUSION: AN UNANSWERABLE QUESTION 242

BIBLIOGRAPHY 245

PREFACE

This piece of work is about why new things happen. The discussion is built around the following three central suggestions:

1. **Novelty is the existence of discontinuity within the distribution of cause-effect relationships over time. A particular novelty is the first instance of operation of a specific, subsequently repetitive, cause-effect relationship.** Novelty is thus the onset of consistent, repetitive, associations between particular changes, from an initial background state in which such changes were either not associated with each other, or were associated only inconsistently (i.e. non-repetitively) over time.
2. **Novelty results from the interaction between distinct systems with similar function but different structure. This is very close to saying that nature is inherently symbiotic, in the widest possible sense.** A system is a structure-function relationship. Any given system exhibits, with respect to its own spatial/temporal boundaries, its own internal/localized pattern of cause and effect (the system's function), and is connected to the outside by a wider, global pattern of cause and effect (the system's structure). When two functionally similar but structurally different systems interact, the structural difference causes temporally inconsistent associations to occur between various changes associated with the different systems. Some of these initially inconsistent associations are subsequently made consistent/repetitive by the interactive response of the functionally similar properties of the two systems. This results in the emergence of a new system, with a structure that is the sum of the starting two systems, but a function that includes repetitive causal rules that did not exist prior to the interaction.
3. **Novelty necessitates that the rules of meaningful language continuously evolve over time and can therefore never have a describable foundation, and explains why time is such a special part of our subjectivity.** Novelty provides an explanation for the disconnect that sometimes exists between what we can think and what we can say - i.e. for the fact that there is an aspect of our thought that is ineffable. The reality of novelty in the world also causes the position of the boundary separating what is, and is not, potentially describable by meaningful language, to evolve over time. Meaningful language thus needs to continually evolve in order to catch up with novelty, the conventions by which language-world homology produces meaningful statements cannot remain fixed in time. Novelty adds new thoughts to our consciousness, initially in an entirely subjective form that is not amenable to objective/meaningful expression in language. This subjective novelty within consciousness is the most direct manner in which we experience the reality of time. This temporality, which is inherent to our subjectivity, is what makes time different from other equivalently

foundational aspects of our experience.

It will always be possible to dismiss novelty as illusory - by claiming that the new events that happen were written into the fabric of existence from the start, by virtue of some fixed, transcendental foundation upon which all of existence is based. Such a foundation would need to be, conveniently enough, outside of the experienced part of the totality of existence and therefore unknowable by definition. But this sort of dismissal is merely a relabeling exercise that defines novelty out of existence so as to avoid having to deal with it. One can be skeptical towards the reality of novelty if one chooses, and argue that although novelty appears real within our experience, it is not actually a real part of existence. But this skepticism amounts to inventing a distinction between experience and existence for which there is no conceptual or pragmatic justification. What cannot be disputed is that the world does seem very much as if it has novelty as an integral part of it: every time we can usefully, meaningfully, speak about a system as having some tangible existence, we can identify, or pragmatically infer with some reasonable level of confidence, a point in time at which that system first came to be.

The scientific method is based on exploiting the continuity, the repetitiveness, of the distribution of cause and effect over time. Science works because when one conducts an experiment on a system, one gets the same result today as one did yesterday. This, by definition, cannot be assumed of systems exhibiting novelty. Therefore, novelty cannot be *fully* understood within the scope of the scientific method. But it is the application of the scientific method to the past that provides the strongest evidence for the reality of the new. Any sort of pragmatic "leap of faith" into a 21st century scientific-secular world view gives us innumerable examples of systems that we are confident genuinely exist now, but did not always exist. Often historical science also allows us to make a decent guess at the point in time at which such systems were formed. More subjectively, in my view, it is self-evident that Earth is the most novel planet, life the most novel natural phenomenon, mind the most novel property of life, humans the most novel species. The history of coevolution between life and the climate of the Earth shows us the kind of contexts in which genuine novelty occurs; and that there exists a tie between climatic extremes and the evolution of novel instances of biological cooperation.

With a brief comparative reading of the respective philosophies of Wittgenstein and Heidegger, I suggest that the concept of novelty sheds light on the relationship between meaningful language and the inherent temporality of human subjectivity. Without novelty, if a question can be asked, it is either answerable (at least in some potential or hypothetical situation), or it is meaningless. If this dichotomy holds, then all that philosophy can ever expect to do is sort out the difference. But with novelty (as opposed to change), what can be asked, what can be answered, and the extent to which these two categories overlap, itself changes over time. This means that the idea of a clean dichotomy between answerability and meaninglessness becomes overly simplistic. This, in turn, goes hand in hand with acknowledging the existence of a similar temporal evolution to the overlap between what can be said and what can be thought. Perhaps, due the reality of novelty, a possibility of genuine discovery from metaphysical speculation may arise, connected to this sort of shifting overlap between private subjectivity and linguistically expressible objectivity.

It is very important to note that, ultimately, what follows is merely my subjective perspective. There is no such thing as an expert in novelty, there never will be, and the reader should be highly skeptical of anyone claiming to be such. This is because in order to become an expert in anything one must become extremely familiar with that thing, which in turn requires that the thing in question be repetitive - and therefore not novel. In contemporary marketing/consumer-driven culture, there exists a relentless pressure to sell one's point of view aggressively. Such self-marketing produces absurdly over confident claims, concerning matters with respect to which the only reasonable response is doubt. The arguments in this piece of work are not, and can never be, certain. Indeed, novelty is the subject area in which claims of certainty are least plausible. Nevertheless, I argue that novelty is, for this very reason, an extremely important concept, and one that we must confront. It seems to me that the above three basic statements, and the arguments related to them, do hold true; the following pages constitute an attempt to explain how and why I consider this to be the case. I ask the reader to accept these suggestions in the speculative, exploratory spirit in which they are intended. This piece of work is aimed at any open-minded person, merely in the hope that they find the discussion thought provoking.

Richard Boyle, 2015

ACKNOWLEDGMENTS

This piece of work developed during the course of several years, whilst I was employed as a postdoc in the Earth system science research groups of Professors Tim Lenton and Donald Canfield. I would like to thank both Tim and Don for their kindness and encouragement, and for cultivating friendly and open-minded research groups. In this regard I thank all members of both the Earth system science group at Exeter university, UK, and the NordCEE group at the University of Southern Denmark. I am also very grateful also thank the U.K.'s National Environment Research Council and Denmark's Nordic Centre for Earth Evolution for funding my Earth system modeling work. I would like to thank the various people who have taken an interest in this piece of work or provided much needed words of encouragement during the period within which it was written, particularly Tais Dahl, Daniel Mills, Andrew Watson, Emma Hammarlund, Nicole Posth, Luke Mander, James Clark, Antonios Karatzas, Leanne Adams, Simon Lanceron, Jennifer Hoyal-Cuthill, Michael Magee, Lamprini Basdeki, Eliza Patouris, Michele Jaret, Louise Bezuidenhout, Ann-Sophie Barwich, Rosalie Tostevin, Christine Wren, Sophie Laursen, Hal Stockley and Kevin Pearson. I thank Holly Buchanan, Ashley Bonem and Bethany Davis of University Press of America for their help during the final stages of the manuscript's preparation. I thank my family for ongoing moral support.

CHAPTER SUMMARIES

The following paragraphs are intended to provide a concise, minimalistic summary of each section of this work. The first chapter introduces the history of thinking about the new, and attempts to summarize the various conceptual impasses that the concept of novelty still raises. The second chapter develops an internally consistent labeling system in order to describe and explain the sequence of events that are necessary and sufficient for novelty, in an attempt to provide what I see as the most general possible perspective from which newness can be understood. The third chapter provides various examples of conceptual problems raised by existing evidence surrounding the history of Earth and life, and argues that appreciation of the reality of novelty may help bypass many such problems. The fourth chapter provides a reading of the philosophies of Ludwig Wittgenstein and Martin Heidegger, and argues that novelty can tie together the concepts of meaningful language and the inherent temporality of our existence, which are raised by the works of Wittgenstein and Heidegger respectively. Finally, it is concluded that although asking "Why does novelty exist?" cannot, at present, be answered in a clear or meaningful sense, the question cannot be dismissed because it reflects an important, unexplained property of the world.

Prelude: Recurrence and recombination

Novelty is a problematic concept for the development of any consistent system of thought, because such consistency is undermined if the rules by which a system is defined can potentially change. Novelty has therefore been defined out of existence by western philosophy since Parmenides. The world's novelty, which our experience suggests to us is real (or at the very least a tangible, realistic possibility), has tended to be attributed to the combined operation of recurrence and recombination. Recurrence and recombination together, when operating within a fixed totality that constitutes all of existence, can conceptually explain why new things appear to happen, and why systems appear to have origins. But ultimately both recurrence and recombination are mechanisms by virtue of which novelty can be dismissed as apparent and illusory. Recurrence relabels novelty as repetition that has not previously been encountered, implicitly due to our experiencing only a restricted proportion of time. We are supposedly confined to one recurrent cycle, but if we were able to see many such cycles we would realize that novelty is in fact repetition over a temporal scale with which we are not familiar. Thus "novelty by recurrence" is illusory because our perspective is short sighted.

Recombination relabels novelty as the reorganization, relative to each other, of permanent unchanging elements of existence. These elements do not themselves originate from anywhere; unless one considers existence as a whole to have an origin, within some singular "first cause"-like act of novelty, a concept of questionable meaningfulness in itself. Thus "recombinatory novelty" is again illusory, it is "just" the emergence of new relationships between things that were always there. A consistent tension has emerged within philosophy between those who consider recurrence and recombination insufficient to explain the novelty within experience, and those who not only regard these mechanisms as adequate, but often question novelty's reality altogether. The latter school of thought has tended to view the concept of novelty and its advocates with suspicion, and sometimes outright derision. It is telling that neither school of thought disputes that novelty appears, at least superficially, very real. Whitehead's emphasis on a process of unfolding of sequences of events towards creative novelty, such that the stuff of existence is events in time, rather than "things" within matter and space, allows novelty to be viewed as metaphysically real. But Whitehead's process philosophy relies on God as the atemporal entity that ultimately initiated this creative unfolding. This emphasis is of questionable explanatory value, because it lumps all the conceptual uncertainties surrounding novelty into an entity (i.e. God) that is, by definition, beyond any meaningful notion of causation or explanation. It is necessary to develop a way of thinking about novelty that includes both this sort of unfolding temporality, aswell as a "common sense" notion of cause and effect within matter and space. It is suggested that in order to more accurately describe and explain novelty, it is useful and appropriate to combine recurrence and recombination with some sort of embodied temporality within the elements that are recombining. This can be done by incorporating into our thinking the notions of structure and function within contextualized systems of cause and effect.

A formal abstract scheme for novelty

Changes are aspects of experience that simultaneously exhibit both identity and temporality. Causal rules are connections between particular changes - descriptions of how and why one particular change precedes another. Causal rules may be repetitive (consistently distributed over time, and therefore amenable to induction-based reasoning systems like science) or chaotic (inconsistently distributed over time and therefore essentially unknowable, although not necessarily non-existent). During an instance of novelty, a causal rule that was previously inconsistent (or non-existent) becomes repetitive. Thus, novelty is the discontinuity that exists within the distribution of causal rules over time, whereas repetition is the continuity that also exists within this distribution. Causal rules are inherently spatially and temporally contextual. A system is a localized, contextual collection of repetitive causal rules. The causal rules within a system can be categorized as either structure or function with respect to the spatial/temporal boundaries of that system. Structural causal rules are global (connect the system to the outside), consistent (temporally invariant), and act as a reference frame. Functional causal rules are local (confined to the system), cyclical (temporally variant) and act as a system-specific network of causes and effects. Novelty results when distinct systems with similar function but different structure interact. The functional similarity causes the interaction to persist over time, and the operation of the causal rules of each system to fall into synchrony with each other. The structurally different nature of the other system acts as a random input, to which each distinct system responds. Crucially, the function-function interaction between the two systems responds to the random nature of the structure-structure relationship, in such a way that initially random changes become "assimilated." During this assimilation, changes that were previously inconsistently associated with each other (due to the random nature of the structural interaction), become consistently associated with each other. This amounts to the origin of new, repetitive causal rules, and the emergence of a new structure-function relationship at the level of the two initial systems combined, in other words, the emergence of a novel system.

Theoretical considerations relating to real instances of novelty during the history of life on Earth

Life is nature's most novel process. This novelty arises from an inherent duality within living systems between information transfer and homeostatic control. It is notable that information and homeostasis refer to two theoretical categories that themselves exhibit a degree of functional overlap, but that manifest in structurally different ways within real systems. Life's duality between information and homeostasis is therefore broadly in keeping with the general principle of novelty by combination between systems with similar function but different structure. Various examples of this duality can be cited within important evolutionary chicken-egg problems that require combination and interaction between two qualitatively different systems, such as the probable requirement for a duality between protein and nucleic acid at the origin of life.

Life introduces a variability to the global biogeochemical cycles of planet Earth that is not present on comparable lifeless planets. It is, however, uncertain whether life regulates Earth's climate (Lovelock's Gaia hypothesis), because what such regulation would amount to is poorly defined. If, however, there is any property of life that biases towards some sort of global climatic homeostasis it is symbiosis. Homeostasis requires the overlap of two "reins" that will pull a system back when a given property of that system deviates either above or below a specific optimum value/range, one rein correcting high, one low, deviations from the optimum range. Sustainable homeostasis also requires that each rein operates within this same optimum range of the variable being regulated. The tendency of symbiosis to fuse together qualitatively different physiologies with shared evolutionary interests but qualitatively different physiologies, amounts to setting up such a system of complementary and compatible homeostatic influences within the life-environment interaction. The potential need for long term climatic homeostasis, in order for life to persist over geologic timescales, means that Lovelock is probably correct in asserting that life is a planetary phenomenon. If it turns out to be true that life also "keeps the Earth in a state that is good for life", this will be due to the generation of physiological novelty by symbioses, and the creation, by such novelty, of homeostatic feedbacks within the life-environment interaction.

Palaeontology, geology and Earth system science have shown that life has a history, during which descent with modification, in conjunction with natural selection, has clearly occurred. But serious conceptual problems emerge with attributing the entirety of the origin of heritable variation in fitness (prior to the operation of natural selection) to random changes in DNA sequences. It is suggested that interactive recombination between qualitatively different biological systems may partly help resolve this conceptual impasse. It is further suggested that the evolution of cooperative complexity is promoted by the combination of extreme climatic feedbacks involving biologically suboptimal environmental conditions, and biological populations with the potential to produce and assimilate such complexity in response to such feedbacks. C.H. Waddington's genetic assimilation mechanism may provide a general means by which to explain the connection between the environment's requirements and the heritable variation in fitness that is generated by life. This amounts to the suggestion that the evolutionary function of DNA is not so much to generate novelty, as to assimilate the novelty that is generated physiologically. The synergy between systems with qualitative structural differences but the underlying functional potential to interact, is likely a useful means by which to explain the history of novelty generation within the natural world.

Novelty, meaningful language, subjectivity and time: A comparative reading of selected works by Ludwig Wittgenstein and Martin Heidegger

The perspective held within the analytic school of philosophy, which was inspired in large part by Wittgenstein's early work, can be represented by the statement that "If a question can be asked, then either this question can be answered (at least in some hypothetical/imaginable situation), or this question is meaningless." This is a statement about which Heidegger, and the continental school of thought to which his work is connected, would adopt a view that is more nuanced and more ambiguous. The concept of novelty provides a perspective by virtue of which both attitudes are correct and complementary - because the statement is correct, but which questions are, and are not, meaningless, changes over time. Novelty in both the world and in our subjectivity necessitates that the rules of logical language evolve over the course of time. Consequently a question that started out as meaningless can subsequently become answerable and meaningful. The language-world homology that separates sense and nonsense cannot itself be described in language (as Wittgenstein said). But this is because the nature of this homology is itself a function of time, due to the reality of novelty (in keeping with the temporality emphasized by Heidegger). The role of novelty within our subjectivity explains why we are, in some sense, embodied time; the core underlying theme of Heidegger's work. The experience of subjective novelty within consciousness is in some ways close to what Heidegger called the "moment of vision", and an instance of novelty in the world is very close to his disclosive truth-concept aletheia. A concept of truth as disclosure, as opposed to convergence, is more compatible with a temporally evolving truth that permits the reality of the new. The fact that we experience subjective novelty within our consciousness is what gives time its directionality, what distinguishes time from the other foundational aspects of our subjectivity. Novel ideas and thoughts enter into our consciousness in an initially wholly subjective form. Analytical thought must "do work" upon such subjective novelties in order to make them objectively communicable in meaningful language.

A central question surrounding the interpretation of Wittgenstein's Tractatus is that of whether the boundary between meaningful language and nonsense is discrete or continuous, i.e. whether there is more than one kind of nonsense, and whether or not an ineffable but metaphysically important category can be postulated. These issues arguably boil down to the question of whether or not we can think what we cannot express in meaningful language. It is suggested that we can indeed think things that are initially incommunicable linguistically, precisely because novelty enters our consciousness in initially inexpressible form. The fact that our consciousness relentlessly generates new thoughts, feelings, ideas etc, necessarily causes our minds to step outside the boundaries of objective language. This is not problematic, because the rules of meaningful language evolve, so as to catch up and recover meaningfulness (until the next instance of novelty, at which point the process of linguistic evolution starts again). The relationship between expressibility (a thought's potential for representation in language) and thought (internal consistency within consciousness) allows the labeling of four categories: Meaningfulness refers to those areas of our thought that we can express in useful language, according to the rules of language games. Nonsense is that which we can express in language, but which does not correspond to consistent thought. Gibberish is that which neither corresponds to consistent thought nor to the rules of expressibility within meaningful language. Subjectivity is that which we can think, and which has internal consistency within our own consciousness, but which we are unable to express within the confines of objectively defined linguistic rules. Novelty enters our consciousness as subjectivity, and turning subjectivity into meaningfulness is a crucial part of both analytical thought, and the natural temporal evolution of interactive language games.

Conclusion: An unanswerable question

A definitive example of a circumstance in which we find our subjectivity "wanting" to step outside the boundaries of meaningful language, is our consistent predisposition to ask (implicitly or explicitly) the question "Why does novelty exist?" In the sense of this work, this question can be posed as "Why is nature predisposed to causing the interaction between systems with different structures but similar functions?" Previous religiously motivated answers to this question have now become meaningless. We realize that asking "Why does novelty exist?" necessitates some violation of the rules of objective language, because there is no possibility of meeting this "Why?" with anything except subjective assertion that steps outside the boundaries of what can be experienced or objectively described. But we nevertheless still feel the inclination to ask the question. The fact that the question is unanswerable does not necessarily demand that we ignore it, because the natural temporal evolution of meaningful language means that it need not remain unanswerable forever.

PRELUDE: RECURRENCE AND RECOMBINATION

"*...there came to be but two ways around philosophy's foundational skepticism about the very possibility of novelty. One of these, recurrence, has the advantage of seeming to have the sanction of nature but the disadvantage of not seeming to offer any real novelty. The other, recombination, seems to offer unlimited novelty, but only if unprecedented relations between existing elements can be considered truly new entities. Despite the equivocal nature of these models, between them they can account for virtually every one of the ways in which novelty has been conceptualized in European history - reformation, renaissance, revolution, invention...*"[1]

Michael North, 2013

Western thought is built upon a foundation in which novelty is implicitly the opposite of rigour. The need for consistency within any system of thought necessitates a ban on new things just coming into that system and changing the rules. It is presumably as a result of this conflict with internal consistency that the ancient Greeks, from Parmenides onwards, simply defined novelty out of existence[2]. It is, according to Gadamer, "*the highest principle of our orientation in the world of experience*"[3] that "*Ex nihilo nihil fit*"; that nothing comes from nothing. Without this principle, we cannot assume the existence of the consistent reference point by which we define concepts. Creating and applying such definitions is, of course, something that we need to do in order to make sense of our experience. If we allow novelty to become part of our concept formation, the meaning of words changes too rapidly for discussions to remain coherent, and we become lost. This pragmatic necessity to exclude the new explains not only why Parmenides imposed "*a sweeping ban on all coming to be*"[4], but also why no subsequent Greek philosopher dared to overturn it.

But novelty is real. One way or another we experience the existence of things that appear, pragmatically at least, to have not always been there. Our investigations into the very systems that we initially used as a benchmark for permanence force us to pragmatically infer the existence of some point in time at which these systems came to be. Most obviously, the author of Ecclesiastes declared with air of authority that "*there is nothing new under the sun.*"[5] But modern science tells us that the radioactive decay of uranium to lead, within pieces of asteroid that formed from "*condensation from a protosolar dust-gas disk*"[6] proves that the origin of the sun itself occurred within "*a brief formation interval corresponding to an age of* 4567.30 ± 0.16 *million years (My).*"[7] Therefore the timelessness of which the Bible speaks must itself have had an origin, about four and a half billion years ago. Declaring the age of a rock, using known radioactive decay rates, perhaps lacks the punchiness and self-important confidence possessed by the authors of the Old Testament. But such scientific investigations have the advantage of a quantitative grounding in the nuts and bolts of the matter around us, aswell, in theory, as the inbuilt capacity for scientific self-correction, should new information come to light. Science works today, so it stands to reason that it should have worked in the past. This applicability of the scientific method to the past is all that one needs to assume in order to be compelled to accept the fact novelty is a ubiquitous and real feature of nature. Overwhelming scientific evidence supports not only a quantitatively definable age for the sun, but also an expanding universe[8], descent with modification during the history of life[9,10], revolutionary and (geologically sudden) transitions in the history of the Earth's climate system[11], a directional (although, of course, not progressive) human evolution from hominids to societies[12] and (much more controversially) maybe even a habitual time dependence in the very laws of nature themselves[13]. For virtually anything that we can meaningfully and usefully describe as a system, we can also infer an origin event for that system at some tangible point in the past. There is therefore no getting away from the reality of the new, once one accepts that science, one way or another, does uncover something approaching the truth, to an extent unprecedented in human history. Confronting this realization, and thereby rejecting any religious or philosophical world view founded upon timelessness, might even be worth considering as the epistemological starting point of the 21st century, science-focused secular age.

The ancient Greeks derived cunning ways by which they could build into their thinking the inconvenient but ever-present appearance of newness in the world. In his (excellent) review of the history of novelty as a concept, North[14] notes how humanity has only ever managed to come up with two mechanisms by which novelty can come into our experience; recurrence and recombination. Recurrence is repetition at the greatest possible scale. When something new happens it is the manifestation of an underlying cycle, through which existence loops infinitely. A recursive procedure is one that is applied once, then applied again to the result of the first application, then applied a third time to the result of the second, and so on. If the result of a sequence of repetitive applications of a recursive operation is that the output generated is the initial input with which the whole process started, then one can speak of a recurrent cycle.

For instance, if one were to generate a sequence of numbers, starting at zero, then adding an even fraction less than one (say a quarter) at each step, this would be a recursive sequence, where the number x at the nth point in the cycle would be related to the previous number at the $n-1$th step by $x_{(n)} = x_{(n-1)} + \frac{1}{4}$. If one then used this sequence and the *sine* function (from trigonometry) to draw a graph with $sine(x \cdot \pi)$ on the vertical axis and x on the horizontal, one would get a repetitively oscillating wave, where one would observe $sine(x \cdot \pi) = 0$ at every fourth value of x. This is because the *sine* function is defined as a repeating wave, such that a graph of $sine(x \cdot \pi)$ goes up to $sine(x \cdot \pi) = 1$ every time that x is halfway between two whole numbers, and falls back to $sine(x \cdot \pi) = 0$ every time x is equal to a whole number. The idea of novelty by recurrence is analogous to the suggestion that time is like the progressively increasing variable x, and that what actually happens within our experience is like the repetitive wave of $sine(x \cdot \pi)$. Ideas centered around novelty by recurrence argue that time as we experience it is restricted. Our time is confined to a single period of the wave, our experience's x value is limited between two whole numbers. When we see what looks to us like a novelty, i.e. we see the graph of $sine(x \cdot \pi)$ take a value that we have never seen it take before, it is not because this novelty is actually real, it is because our perspective is restricted. We are stuck within only one iteration of the underlying repetitive cycles of the universe. Our time is stuck between (say) $0 \leq x \leq 1$, so although the value of $sine(x \cdot \pi)$ that we experience on our graph is a "novel" value to us, because we have never previously seen it, if we were a little wiser, and our temporal frame of reference was extended to (say) $0 \leq x \leq 10$ and all the x values in between, we would see that what we thought was a genuine novelty is in fact only the universe's inherent cycle repeating itself.

Parmenides established this sort of notion of such an underlying repetitive cycle, Empedocles, Plato and Aristotle left it in place. This ancient Greek idea that all experience is ultimately being driven by the repetition of an underlying cycle of recurrence persisted; within the notion of the Christian Baptism as a re-enactment of the creation of man[15], through Nietzche's belief in the eternal recurrence[16], all the way to Roger Penrose's cosmological models - in which the universe undergoes repeating cycles of "big bang" and "big crunch" events[17]. Recurrence provides a system within which one can explain away the appearance of the new by arguing that time, as we experience it, is restricted to an artificially narrow interval. If we experienced time as it really is, we would see that the novelty that appears to be there is, in fact, illusory.

But even if we allow ourselves to believe that there does exist a foundational cycle of recurrence at the base of existence, and this cycle gives our experience of existence the (illusory) appearance of newness, then we are still left with the question of "How?." Something is present at one point in time that was not present at the previous point. It seems new to us, within our restricted part of existence, even if it is not genuinely new to existence as a totality. What has actually happened between the first time point and the second? How does the recurring loop give us the impression (accurate or not) of novelty? This question presents to us the other central feature of all previous conceptual frameworks for the new: recombination. The fixed nature of being is a totality, but this totality consists of variable relative arrangements of being's most elemental constituent properties. The recombination of these different constituents, and the change in their relative arrangement to each other, is what accounts for "new" things happening. One cannot change the nature and total amount of stuff that there is, but one can change the way in which the constituents of existence are organized relative to each other, the way in which they are recombined.

As with recurrence, the concept of recombination dates back to Parmenides:

> "*In his poem 'On Nature' a Goddess instructs him that reality must necessarily be, or must necessarily not be, or must both be and not be, which is impossible. Given the first option, it can be deduced that what is real must be ungenerated, imperishable, indivisible, perfect and motionless. This, the Parmenidean One, obviously contrasts with the relative and specious appearances of things, which arise only through the opposition of two equally unreal forms, Light and Dark...*"[18]

The same principle, of changing relative proportions of the elements of a fixed totality, was elaborated upon by all subsequent Greek philosophers, but never challenged. For Empedocles, the changes of experience, and the particulars into which reality is divided, are ultimately the result of changes in the relative dominance of love and strife:

> "*There are these alone; but, running through one another, they become men and the tribes of beasts. At one time they are all brought together into one order by Love; at another, they are carried each in different directions by the repulsion of Strife, till they grow once more into one and are wholly subdued. Thus in so far as they are wont to grow into one out of many, and again divided become more than one, so far they come into being and their life is not lasting; but insofar as they never cease changing continually, so far are they evermore, immovable in the circle.*"[19]

Having defined the real as an inescapable totality "*that which keeps its own form unchangingly, which has not been brought into being and which is not destroyed, which neither receives into itself from anywhere else, nor itself enters in to anything else anywhere.*"[20], Plato followed in the recombinatory tradition, explaining away the apparent novelty of existence by means of variability in the relative arrangements of air, earth, wind and fire. These traditional elements he in turn decomposed into two sorts of triangles, apparently as a result of the attitude that "*As the irreducibly basic representation of a plane, the triangle can be thought of as the unit of three dimensional reality, the building block which in various combinations makes up everything we see*"[21]. In this vein, the ancient Greek tradition of a glaring lack of imaginative novelty, within the systems of thought devised to cope with the novelty apparent in nature, continued to Aristotle "*there is always something which underlies what comes to be, from which what comes to be comes, for instance animals come from seed*"[22]and so on. Reality, existence, was an irreducible, fixed whole, yet it had the appearance of being divisible into element-like constituents. Different combinations of these constituents gave rise to the appearance of things that were not previously there. Some aspects of these reorganizations were cyclical, recurrent- there existed a time component to the sort of recombination happening. This explained the evolution of nature and the temporal sequences of distinct reorganizations of the stuff of reality. By combining the concepts of recurrence and recombination, the Greeks were able to explain away any and every aspect of novelty, whilst avoiding having to admit that it might be real.

The tradition of novelty by recombination and/or recurrence persisted up to the present day with its foundations largely unmodified. The Neo-Darwinian modern synthesis (in which Darwin's natural selection concept was connected to Mendel's evidence for particulate inheritance, by the statistician R.A. Fisher)[23] was the crowning glory of theoretical attempts to ground novelty in recombination. Genetic combinations produce all the novelty of the living world, survival of the fittest does the rest. Even Bergson, who rebelled against this evolutionary orthodoxy, by arguing for the existence of a qualitative novelty that could only be explained by some form of "vital force" distinguishing life from non-life, still did not deviate from the paradigmatic notions of recombination and recurrence. Bergson[24–26] argued that the creativity manifest in the history of life amounts to the unfolding of a continuous program, such that the causal elements of evolution do not so much combine with each other, as spread out like a wave, releasing their potential creative power. He suggested that the instincts of animals are directly homologous to the "mystical intuition" of man, that duration, the cumulative experience of events, is qualitatively different from the concept of physics-time as a "normal" dimension akin to space, and that memory is inherent in nature and nature's laws are therefore habitual rather than fixed. Bergson's philosophy therefore asserts the existence of a deep and continuous creativity in nature, aswell as a equally deep and continuous memory of the past within time as we experience. But Bergson remained vague about exactly how this creative force was supposed to work. One gets the impression it is along the lines of novelty by recombination in reverse - such that the elements of existence spread out and release their potential to do something new once they are uncoupled from that to which they were previously joined.

But Bertrand Russell was unconvinced that Bergson had added anything to ancient Greek thinking, and was uneasy with Bergson's artistic style, for example describing how in Bergon's writing:

> "*..there is the great climax in which life is compared to a cavalry charge: 'All organized beings, from the humblest to the highest, from the first origins of life to the time in which we are, and in all places as in all times, do but evidence a single impulsion, the inverse of the movement of matter, and in itself indivisible. All the living hold together, and all yield to the same tremendous push. The animal takes its stand on the plant, man best rides animality, and the whole of humanity, in space and in time, is one immense army galloping beside and before and behind each of us in an overwhelming charge able to beat down every resistance and to clear many obstacles, perhaps even death.' But a cool critic, who feels himself a mere spectator, perhaps an unsympathetic spectator, of the charge in which*

man is mounted upon animality, may be inclined to think that calm and careful thought is hardly compatible with this form of exercise."[27]

In another, later critique of Bergson's work, Russell goes on to highlight how the problem of explaining the new was, in the early 20th century, essentially unaltered from the form that it had taken for the ancients. Russell was obviously irritated and unconvinced by Bergson's claim to have surpassed inductive reasoning through some underlying principle of creativity:

> "*If one might venture to apply to Bergson's philosophy so vulgar a thing as logic, certain difficulties would appear in this philosophy of change. Bergson is never tired of pouring scorn upon the mathematician for regarding time as a series, whose parts are mutually external. But if there is indeed genuine novelty in the world, as he insists (and without this feature his philosophy loses its attractive qualities) and if whatever really comes into the world persists (which is the simple excess of his doctrine of duration), then the sum total of existence at any earlier time is part of the total at any later time. Total states of the world at various times form a series in virtue of this relation of whole and part, and this series has all the properties that the mathematician wants and that Bergson professes to have banished. If the new elements which are added in later states of the world are not external to the old elements, then there is no genuine novelty, creative evolution has created nothing, and we are back in the system of Plotinus. Of course Bergson's answer to this dilemma is that what happens is 'growth', in which everything changes and yet remains the same. This conception is, however, a mystery, which the profane cannot hope to fathom. At bottom, Bergson's appeal is to mystical faith, not to reason; but into the regions where faith is above logic we cannot follow him."*[28]

Whitehead confronted the difficulties involved in formulating the problem of novelty in a clearer and more direct manner than Bergson. Whitehead's process philosophy[27] is based on the idea that the foundational elements of reality are "actual entities." All actual entities are "occasions of experience", which (crudely speaking) amount to sequences of events. Thus the pieces that make up the stuff of reality are temporal processes, rather than material elements, Platonic forms or Aristotelian substances. The fundamental foundation of existence is not matter within space, or ordered sets of rules, it is the unfolding of processes through sequences of events. There is one exception to this statement, one actual entity within Whitehead's thought that is simultaneously atemporal, aswell as (like the other actual entities) temporal. This unique actual entity sets up in advance the whole process of unfolding, and it is this process that relates all the other actual entities in a continuous process of becoming. This uniquely atemporal actual entity is God. Whitehead explicitly asserts the reality of creativity. The actual existence of actual entities is a continuous process of becoming and states that "*'becoming' is a creative advance into novelty....*" Thus, for Whitehead, the ultimate constituents of reality are fundamentally temporal (e.g. a human being is a sequence of many different "occasions of experience", rather than a physical body or a transcendental soul). The efficient/singular causes that common sense might have us believe underlie material events are supplemented by "nomic" lawlike temporal regularities - in which causation is more network-like than linear. For Whitehead:

> "*the fundamental fact of cosmology is the unfolding of processes, not conceived of in terms of modifications of underlying substances but united by networks of relations, and selected in advance by God. A process theory of mind sees its unity not in terms of underlying substance, but as the relatively permanent form or pattern displayed by continuously changing processes, just as a wave maintains its form although it is constituted by different volumes of water at different times.*"[29]

In other words, things are temporal more than they are either material or transcendental, novelty arises by recombination of overlapping flows of event sequences, rather than interactions between particles, and this overlapping is set in place by God. Thus Whitehead introduces the flow of time into recombinatory novelty- and this introduction is his major addition to human thought. But he still relies on recombination, and he explains its occurrence by virtue of an appeal to the existence of God. In this sense, as North puts it, the philosophies of Whitehead and Bergson "*are really additions to the history of the new and not accounts of it, except insofar as [they] identify novelty as one of the great unsolved problems in human thought.*"[30]

Terrence McKenna added a poetic information-age spin to Bergson's emphasis on the directionality of time, and the distinction between events of experience and the homogeneity of "physics time." McKenna's "Timewave zero novelty theory"[31] essentially argues that the "rate of history", in terms of the accumulation of new events and complexity, is somehow accelerating, because history is being "pulled" by some sort of discrete and special event in the future. This event, a transcendental object at the end of time, a teleological attractor within the future, was (supposedly) a means by which the universe would end in some sort of singularity of infinite complexity. McKenna claimed to have found a recursive pattern, implying accelerating complexity towards such a singularity in the ancient Chinese scripture the I Ching[29]. He viewed novelty as progressive and non-linear, giving rise to order and complexity that counteracted the inherent habitual and repetitive properties of nature. Serious assessment of McKenna's thought is rendered somewhat problematic by the fact that his prediction of a 2012 Mayan-apocalypse infinite novelty event seems to have failed to transpire. Even letting this pass, the vague, imprecise nature of McKenna's definition of novelty is what clouds and confuses assessment of his theories. For instance, he appears to have regarded novelty and complexity as synonymous, interchangeable terms, without clearly explaining why. McKenna's key conceptual addition to Whitehead's process theme was the idea of the finite nature of human history, manifesting in an infinitely complex future event that somehow pulls the past and present towards it. The case for the existence of such an event seems confused and vague. Nevertheless, Terrence McKenna does deserves credit for being the first to explicitly propose that humanity is nature's most novel species[30], and his arguments about human technology's accelerating drive towards increasing complexity surely have at least a grain of truth to them.

Alain Badiou[31] breaks with Parmenides' foundational emphasis upon the "the one" (i.e. some sort of irreducible totality that bounds all of existence), raising the intriguing possibility of a way out of the trap between recurrence (of a cycle within the one) and recombination (of stuff that manifests during the course of such a cycle). Badiou speaks about creative novelty in terms of the "event", which (paraphrasing) is some of rupture from the norm of what went before and what gave rise to that event's occurrence. For Badiou, for instance "*situations are nothing more, in their being, than pure indifferent multiplicities....*" Badiou states that "*a truth is solely constituted by rupturing with the order which supports it, never as an effect of that order. I have named this type of rupture which opens up truths 'the event'...*" and "*although it is situated in a world, a truth does not retain anything expressible from that situation...*" All of this goes hand in hand with the assertion that essentially any situation contains the potential for innovation. This ubiquitous potential for the new does not conflict (apparently) with a tie between a given event and the onset of adherence to the truth of that event by a subject, a "*militant of truth.*" Thus the continuous potential for novelty is linked with an inherently disclosive, revelatory truth concept. There is some kind of connection between a rupture to novelty, and such a rupture exhibiting an indiscernibility within the situation from which the event stems. In other words, for Badiou, when an event of creative novelty happens, this entails some sort of pointing to the outside by means of some property of that event.

What exactly any of this really means, and why any of Badiou's assertions are supposedly true, is difficult to determine. It seems that in Badiou's thinking there is some sort of connection between the predisposition to novelty and what he calls "*inconsistent multiplicity.*" This is a term borrowed from set theory, and reflects Badiou's belief in some sort of underlying Platonic existence of mathematical entities - that "*mathematics is ontology.*"[31] In Badiou's thought, sets, and the capacity for belonging and inclusion, by which elements may be related to sets, literally have some sort of existence outside the minds of mathematicians. Many of his claims are justified by varying degrees of reference to Cantor's continuum hypothesis (that that there is no set with more elements than the (infinite) set of all natural numbers, but less elements than the (infinite, but larger) set of all real numbers[32]). Exactly why the continuum hypothesis, set theory, and indeed mathematics as a whole is supposed to be relevant to things that aren't mathematics, let alone the foundation of existence, it is impossible to tell from Badiou's writing. This ambiguity arises largely because the justification for Badiou's sweeping claims comes in the form of statements like:

> "*Ontology, axiom system of the particular inconsistency of multiplicities, seizes the in-itself of the multiple by forming into consistency all inconsistency and forming into inconsistency all consistency. It thereby deconstructs any one-effect; it is faithful to the non-being of the one, so as to unfold, without explicit nomination, the regulated game of the multiple such that it is none other than the absolute form of presentation, thus the mode in which being proposes itself to any process.*"[31]

This might, perhaps, mean something like "situations have within them a sort of uncontextualized variety, which they can sometimes (somehow) project onto the outside of themselves. This projection creates (somehow) the potential for this variety to cause something new within that which the situation is initially different from." Or perhaps Badiou means something entirely different, it is impossible to tell with any reliability, because his language is so ridiculously convoluted and obscure. In this vein, Badiou's critics have argued that he extrapolates to situations completely unrelated to set theory (or mathematics) but uses the jargon of set theory as an appeal to authority. For instance, Scruton states that "*the jargon of set theory is waved like a magician's wand, to give authority to bursts of all but unintelligible metaphysics...*"[32] Badiou's mix of set theory and politics raises problems in the minds of the mathematicians themselves; for instance Ricardo and David Nirenberg state that:

> "*Alain Badiou calls himself a Platonist and proclaims the revolutionary political power of his philosophy of numbers. But insofar as his mathematical ontology disguises the contingent in robes of necessity, it can only diminish our freedom. We can embrace the politics if we so wish. But we should not confuse this choice with mathematics, nor can we call it philosophy.*"[33]

With regards to the idea that "mathematics is ontology" and that set theory is somehow relevant to aspects of human life other than set theory, Wittgenstein's philosophy of mathematics comes to mind; "*Our suspicion ought always to be aroused when a proof proves more than its means allow it. Something of this sort might be called a puffed up proof.*"[34]

But perhaps such criticism of Alain Badiou's philosophy is peripheral. Perhaps attacking the obscurant nature of Badiou's style amounts to missing the point, and there is something underneath of profound importance, which his critics have failed to understand. Perhaps the important thing is that a contemporary philosopher has recognized that there is a fundamental problem with the Parmenidean One's failure to incorporate the reality of the new (beyond the continual iteration on the inter-connected themes of recurrence and recombination). Badiou has built his system of thought around the assertion of a universal potential for novelty. Maybe the key thing is the assertion itself, rather than the means by which this assertion is justified (innappropriately de-contextualized mathematics). The natural, productive manner which to build upon Badiou's assertion, is to determine precisely how a particular situation that realizes its potential for novelty differs from one that does not.

Aside from the discussions arising within academic philosophy itself, there are two immediately obvious means by which a skeptical person might dismiss novelty as trivial, or otherwise unworthy of investigation: First, the argument that novelty is the unfolding of a continuous change, a change which gives time a directionality, but which is still nevertheless the unfolding of a constant underlying relationship. But novelty is simply not the same thing as change. Change is one of the most problematic concepts in human thought, but boils down to some kind of combination between identity (there must be a thing changing) and time-ness (the changing thing must have an underlying temporality that allows us to mark it out as a change). Change allows us to see that time is a feature of our existence. But within change, the same underlying relationship folds out as time progresses. Novelty, by contrast, is the origin of qualitatively new relationships. For example, consider entropy. The second law of thermodynamics states that the entropy (disorder or granularity) of the universe can only ever increase[35]. The second law of thermodynamics arises because for any closed system we can distinguish an aggregate macrostate, and observe the number of microstates (combinations of the components of that system) that can give rise to a given macrostate. If we know the underlying mechanisms describing how the components combine to make an aggregate (specifically if we know that molecules diffuse in response to heat so as to become evenly distributed over a given space), we can determine the probability of one macrostate happening rather than another. The increase in entropy within any closed system results from the fact that a granular, evenly distributed macrostate is more probable than a specific, irregularly distributed one. More combinations of individual pieces of stuff are possible when the aggregate arrangement is a diffuse, spread out one, than when the aggregate arrangement is an elaborate, precise, ordered one. The key point is that although the second law of thermodynamics does invoke a directional "arrow of time", this arrow results from the repetition of the same underlying causal principle. So when the entropy of the universe increases, this is an example of change, but change derived from repetition rather than from novelty. By contrast, the origin of the second law of thermodynamics, the putting in to place of a state of affairs that gave rise to the kind of universe in which disorder only ever increases, in which a heat death is inevitable, would be an example of novelty.

The second potential means by which a skeptical person might dismiss novelty as trivial is the anthropic principle. From an anthropic perspective, one might argue that novel events are a prerequisite for the existence of observers, therefore we cannot proceed as if it is surprising, problematic, or confusing that novelty exists, because we, observers, also exist. More generally, the anthropic principle is a method of argument by which some property of the world is "explained" by the observation that if that property did not exist, it would not be possible for us to be here discussing its existence via observations. The modern anthropic principle derives from the fact that astrophysicists noted that the universe's being conducive to the emergence of observers cannot be considered a paradoxical feature, because it is certain that the universe has produced observers, and a universe that failed to do so would, by definition, be unobservable[36]. Blackburn communicates the anthropic principle as analogous to highlighting how misguided it would be to ask "Why do I always see roads when I go driving?"[37]. In my opinion the anthropic principle is best illustrated by a quote from the main character in a novel by Haddon:

> *"People believe in God because the world is very complicated and they think it is very unlikely that anything as complicated as a flying squirrel or the human eye or a brain could happen by chance. But they should think logically and if they thought logically they would see that they can only ask this question because it has already happened and they exist. And there are billions of planets where there is no life, but there is no one on those planets with brains to notice. And it is like if everyone in the world was tossing coins, eventually someone would get* 5698 *heads and think that they were special. But they wouldn't be because there would be millions of people who didn't* 5698 *heads."*[38]

Assuming that novelty is a prerequisite for observers, one might dismiss the question of the new in this way. However, in my view, though nominally valid, the anthropic principle is something of cop-out. The most we can get from viewing the world through the lens of anthropic "observer-bias" is an appreciation of the need to change a "why?" question to a "how?" question. This relabeling exercise teaches us nothing. For instance, given some cosmological or metaphysical constraints, someone might ask "Why is existence this way?", which could legitimately be dismissed with "if it were not, we would not be here to ask." But all that would be gained would be a rephrasing to "Given that existence could conceivably be many other ways, how is it the case that existence is in fact is this (observer-permissable) way?." I.e., we simply rephrase our question to as to ask how our universe differs from the (apparently) infinite number of possible universes that would have failed to produce observers. We ask "how?" instead of "why?" but, beyond this, we do not really get anywhere. To refer to Haddon's metaphor above, notwithstanding the importance of chance, and the effect of large sample sizes in random phenomena, if any one of us literally did sit, alone, throwing an unbiased coin, and ended up with 5698 heads in a row, we would be shocked. We would and (rightly or wrongly) feel compelled to inquire as to whether there was something special about our circumstances. We would not dismiss the occurrence of 5698 sequential heads at a probability of $(\frac{1}{2})^{5698} \simeq 0$ as an artefact of our being only one of millions of other observers from (say) parallel universes or convergent coin experiments. We would investigate whether it was indeed a fair coin. This illustrates how, in practice, the anthropic principle can be misused so as to provide an excuse for avoiding embarking on an investigation. If the universe produces novelty, we should ask what property of the universe it is that creates this disposition, rather than try to convince ourselves that novelty is unremarkable.

Yet in our age, novelty is indeed systematically dismissed as unremarkable. North[39] notes how the very word "novelty" has more pejorative than metaphysical connotations in the modern vernacular. A "novelty" is a tacky gimmick, a vulgar distraction, a gratuitous technological gadget. Metaphysical discussion surrounding the new is perceived as misguided, emotionally driven, and often anti-science and/or religiously motivated. As Stace caricatured the viewpoint of novelty's defenders in the 1930s:

> "*Science finds, or used to find, the world completely governed by law. All events are reduced to cases of causal or functional determination. This means, it is alleged, that there can be no genuine novelty in the world. We shall have a mechanical universe, an eternal repetition of unalterable sequences, the everlasting turning of wheels upon wheels. All change is mere rearrangement of old elements in new patterns. The end is foreseeable in the beginning, is contained in the beginning. The universe cannot produce anything which was not implicitly present from the very beginning, that is, it cannot produce any novelty. The concept of novelty in philosophy is a revolt against this mechanical view of the world which is the product of science. Secondly, it is a revolt based upon an emotional revulsion. The philosophers of novelty dislike the scientific picture. They desire a world in which what they call genuine novelty shall be possible. And because they wish for such a world they attempt to prove that the world really is of this sort.*"[40]

Even if this characterization is correct, one might suggest that the critical attitude suggested by comments of this sort entails an equal level of emotional investment: in setting up a fixed, consistent system of thought in which novelty remains vague and redundant, predictive power reigns supreme, and man has conquered nature. The materialist world view was/is equally committed to hammering in the final nail in the coffin of any theological-philosophical argument for some sort of first-cause[41] origin for existence. Neither perspective seems particularly impartial. Importantly however, it is telling that both pro and anti-novelty thinkers do seem to agree that the world does seem very much like it contains real newness; they merely differ as to whether or not they consider this appearance meaningful or misleading.

The fact that novelty conjures up images of tacky triviality probably has more to do with the unchallenged march of western consumerism than the fact that the philosophical debate is stuck at some interplay between recurrence and recombination. The fact that the political classes who run the west regard profit as an unquestionable law of nature, around which anything and everything must be moulded, has given rise to a marketing-centric bureaucratic culture, in which everyone is constantly required to sell themselves. The open-ended nature of a profit-centric world view creates, in theory, a pressure to strive for innovation. But in practice it results in a system in which everyone relentlessly needs to justify their professional and social existence by pretending to be creative - whilst in fact doing a job that is at best necessary but repetitive, at worst profitable but devoid of any value, creative or otherwise. We have become conditioned to the claims of novelty with which we are constantly bombarded, to the point that we simply do not believe these claims anymore, As Graeber puts it "*a timid, bureaucratic spirit suffuses every aspect of cultural life. It comes festooned in a language of creativity, initiative, and entrepreneurial-ism. But the language is meaningless.*"[42] We are so used to people telling us that things are new, are revolutionary, when they are not, that we fail to notice the genuinely powerful novelty apparent in natural history, and fail to entertain the notion that it might be metaphysically significant. A rare current of rebellion against the prevailing trend of looking at novelty with derision comes from Crosby - who notes that systems of efficient causes require novelty but cannot explain it, and goes on to argue that cultivating creativity over adherence to repetitive rules would lead to a more productive and empowering education system[43]. But such detractors are few and far between. The modern western world constantly tries to convince itself that it values novelty, whilst at the same time shoehorning every specific innovation into the repetitive march of profit, and viewing metaphysical discussion of novelty as regressive.

I consider the reality of novelty indisputable. But I make no attempt to defend this claim, because a dismissive skeptic will always be able to relabel any actual newness as the unfolding of some pre-set causal system or cycle that, although it is outside existence as we experience it, was "obviously" always there. (It is a peculiarity of the secular/materialist age in which we live that such a skeptic would, today, also be able to claim that such an attitude is scientific). The reality of novelty is, in this work, a starting assumption. The question is what it actually means and how it actually happens. Novelty undoubtedly does manifest by virtue of some interaction between the mechanisms of recurrence and recombination. But this is not enough. Whitehead's reorganization of thinking about the new, into the unfolding of a foundational temporality, is a strong positive contribution, but this is not enough either. In this work I attempt to combine the two classes of idea by speaking in terms of basic elements that can be recombined with each other, but giving each element a temporality that is local and specific to it (as opposed to an aspect of some universal flow of time). I suggest that progress can be made by thinking in terms of contextualized systems of cause and effect. Novelty happens when systems with similar function but different structure are forced to interact.

It is a reflection of how foundational the concept of novelty is to how we think that the discussion of the above paragraphs was possible and (hopefully) made sense, despite the fact that I made no attempt to define novelty. As discussed in the preface, the sense in which I use the word here is as a discontinuity in the distribution of cause and effect within time. Cause and effect can be contextualized to "elements" that can be recombined in a conventional way so as to generate novelty, but which simultaneously exhibit an underlying temporality that allows us to regard those elements as systems in their own right. Such systems have a local, repetitive pattern of cause and effect - i.e. a temporality analogous to the actual entities of Whitehead's process philosophy, but a spatial frame of reference to which that temporality is confined. This local temporal repetition, specific to a particular system, I label function, and the spatial frame of reference within which this function is confined, I label structure. In other words, the stuff of existence is indeed distinct elements that can be recombined, but at the same time each of these elements has a temporality (function) that is confined to their local spatial frame of reference (structure). In this way, one can recombine particle like elements of existence, but at the same time one can recombine functional, temporal events. When the functional repetition of two different systems fall into synchrony with each other, and at the same time the structures of two different systems act as a random input with respect to the other, novelty can result. The combination of synchrony and random input causes the emergence, not of new arrangements of the particulate constituents of existence, but of qualitatively new events, new structure-function relationships. This idea will be discussed in more detail in the next chapter.

Notes

1. North, M. *Novelty: A History of the New.* (Chicago, 2013).
2. Waterfield, R. *The first philosophers: The pre-socratics and sophists.* (Oxford, 2009).
3. Gadamer, H.G. *The beginning of philosophy.* (Bloomsbury, 2000).
4. Mourelatos, A.P.D. *The Pre-Socratics: A Collection of Critical Essays.* (Princeton, 2014).
5. *Holy Bible*, English Standard version. (Old Testament composed approximated 3rd century B.C).
6. Amelin, Y., Gosh, A. & Rotenburg, E. *Unraveling the evolution of chondrite parent asteroids by precise U-Pb dating and thermal modeling.* Geochimica Cosmochimica Acta. 69(2). 505 – 518. 2005
7. Connelly, J.N., Bizzarro, M, Krot, A.N., Nordlund, A, Wielandt, D. & Ivanova, M.A. *The Absolute Chronology and Thermal Processing of Solids in the Solar Protoplanetary Disk.* Science 338. 651-655.2012.
8. Kragh, H. *Cosmology and Controversy.* (Princeton, 1996).
9. Darwin, C. *On the origin of species by means of natural selection.* (Murray, 1859).
10. Ridley, M. *Evolution.* (Wiley-Blackwell, 2003).
11. Lenton, T.M. & Watson, A.J. *Revolutions that made the Earth.* (Oxford, 2011).
12. Wood, B. *Human evolution: A very short introduction.* (Oxford, 2006).
13. Sheldrake, R. *A New Science of Life.* (Park Street, 1995).

14. North, M. *Ibid* ref. 1.
15. Ladner, G.B. *The idea of reform: Its impact on Christian thought and action in the age of the fathers.* (Harvard. 1959).
16. Nietzsche, F. *Thus spoke Zarathustra.* (Schmeitzner, 1883).
17. Penrose, R. *Cycles of Time: An extraordinary new view of the universe.* (Vintage, 2012).
18. Blackburn, S. *Oxford dictionary of Philosophy.* (Oxford, 2008).
19. Empedocles 490 – 430 B.C. (Fairbanks, A. *Empedocles: Fragments and Commentary.*1898).
20. Plato. *Timaeus*, Socratic dialogue. ~360B.C.
21. Pellicia, H. 2001. *Selected Dialogues of Plato: The Benjamin Jowett Translation.* (Modern Library Classics, 2001).
22. Sachs, J. *Aristotle's Metaphysics.* (Trade Select, 1999).
23. Ridley, M. *Evolution.* (Wiley-Blackwell, 2003).
24. Bergson, H. *Creative Evolution.* (Forgotten Books, 2015. (First published 1907)).
25. Bergson, H. *Time and Free Will: An essay on the immediate data of consciousness.* (Dover, 2001 (First published1910)).
26. Bergson, H. *Matter and Memory.* (HardPress,2012. (First published1912)).
27. Russell, B. *The Philosophy of Bergson.* (Cambridge, 1914).
28. Russell, B. *Skeptical Essays.* (Routledge, 1928).
29. Whitehead, A.N. *Process and Reality: An essay in cosmology.* (MacMillan, 1929).
30. North, M. *Ibid* ref. 1.
31. McKenna, T. & McKenna, D. *The invisible landscape: Mind, Hallucinogens and the I Ching.* (HarperOne, 1994).
32. Scruton, R. *A Nothing Would do As Well.* (Times Literary Supplement. 31st August 2012).
33. Nirenberg, R.L. & Nirenberg, D. *Badiou's Number: A Critique of Mathematics as Ontology.* (Critical Inquiry 37, University of Chicago. 2011).
34. Wittgenstein, L. *Remarks on the Foundations of Mathematics.* (Wiley Blackwell, 1981(First published 1953)).
35. Atkins, P. *The laws of thermodynamics: A very short introduction.* (Oxford, 2010).
36. Carter, B. *Large Number Coincidences and the Anthropic Principle in Cosmology.* IAU Symposium 63, pp. 291–298. 1974.
37. Blackburn, S. *Ibid* ref. 18.
38. Haddon, M. *The curious incident of the dog in the night time.* (Vintage, 2004).
39. North, M. *Ibid* ref. 1.
40. Stace, W.T. *Novelty, indeterminism and emergence.* The philosophical review 48. 296 – 310. 1939.
41. McInerny, R and O'Callaghan. *J. Saint Thomas Aquinas.* Stanford Encyclopedia of Philosophy. 2015.
42. Graeber, D. *Of flying cars and the declining rate of profit.* The Baffler, 19. 2012.
43. Crosby, D. *Novelty.* (Lexington.2005).

A FORMAL ABSTRACT SCHEME FOR NOVELTY

This section contains a development of the central suggestion; that novelty results when systems that function in similar ways but are composed of different stuff combine and interact in time and space. In an attempt to give a maximally general description, much of this section is an attempt to develop formal and precise definitions for the contextual meaning of words like "structure", "function", "system" , "cause", "effect", etc. After this clarification there follows a description of the sequence of events that is necessary and sufficient to give rise to novelty. So as to make the assumptions as transparent as possible, a few symbols have been borrowed from logic and set theory. However, it should be noted that the purpose of these symbols is to make the assumptions clear, rather than to provide a mathematical or logical "proof" of the argument's truth. The truth (or not) of the argument is something that can only be assessed by individual reflective comparison between this argument and the world. Various conceptual issues from natural sciences, relevant to such a comparison, will be discussed in the subsequent chapter. The purpose of this chapter is just to precisely state what the argument is. The following definitions will be used for the various symbols:

- $\exists$ "there exists" ($\nexists$ "there does not exist").
- $\forall$ "for all."
- $|$ "such that" (e.g. $x|y$ means "x such that y").
- $\in$ "is a member of", e.g. $x \in A$ means thing x is contained in category A, ($\notin$ "is not a member of").
- $\neg$ "negation of", e.g $\neg x$ means "x is untrue."
- $:$ "is a function/rule from" e.g. $f : A \longrightarrow B$ means f is a function (rule, relationship) that relates thing A to thing B. The starting entity A is called the function's input "domain", the finishing entity B the function's output "co-domain."
- $\rightarrow$ "relates to" e.g. $f : A \longrightarrow B$, which can also be written $A \overset{f}{\longrightarrow} B$, means "rule f relates thing A to thing B."
- $\circ$ "the composite of" the combined result of the sequential operation of two (or more) rules, e.g $f \circ g = h$ means "the operation of rule f following the operation of rule g is equivalent to the operation of rule h."
- $\circlearrowleft$ "is an endomap" (a rule that relates a thing to itself), a function whose input ("domain") is the same entity as its output ("co-domain"), e.g. $A^{\circlearrowleft f}$ = "function f relates thing A to itself."
- $\Longrightarrow$ "implies", i.e. $x \Longrightarrow y$ "the truth of x implies the truth of y."
- & "and", i.e. $x\&y$ means "both x and y are true", so $x\&y$ defines that proposition that is only true if both x and y are true.

- $\vee$ "or" i.e. $x \vee y$ means "either x, or y, or both, are true", so $x \vee y$ defines that proposition that is true unless both x and y are false.
- $|S|$ "The size of the set/category S", i.e. the number of individual things/entities in the set/category S.
- S^C "complement." The complement of a set/category S is everything outside, i.e. not a part of, that category S.
- $\emptyset$ the empty set (i.e. the set containing no elements).
- $\cap$ "intersection", e.g. $A \cap B$ is the region of overlap between two sets/categories A and B. Note that for any set S, the intersection between that set and its own complement is empty by definition. By the very nature of sets, no element/entity can simultaneously be contained within a set/category and be outside of that same set/category, i.e. $S \cap S^C = \emptyset, \forall S$; "the intersection between set S and its complement S^C is empty for all possible S's."
- id "identity function", a relation with an output that is the same as its input. $id(x) = x, \forall x$ "the identity function of x is equal to x for all particular x." (An example of an identity function might be simply "multiply by one when x is any number" $id(x) = x * 1 = x, \forall x$, meaning "for all possible x the value of x multiplied by 1 is the same as x").
- $\equiv$ "is equivalent to" (often used when things are conceptually equivalent without being formally, mathematically equal).
- $\cong$ "is congruent with" (has the same structure as, is consistent with).
- $\perp$ "contradiction", e.g. $x \perp y$ means "the truth of x contradicts the truth of y."

Summary in words

A change is any element of experience simultaneously exhibiting both identity and temporality. Causal rules are the connections that exist between particular changes. Causal rules are either consistent or inconsistent in terms of their occurrence over time; i.e. in terms of whether of not they reliably describe the temporal connections between particular changes. Causal rules that operate consistently over time are categorized as repetition. Causal rules that operate inconsistently over time are categorized as chaos. No causal rule is simultaneously chaotic and repetitive. Chaotic causal rules are unknowable to any degree of reliability, whereas repetitive causal rules are amenable to induction-based systems of reasoning such as science. Novelty is the discontinuity that exists in the distribution within time of repetitive causal rules. In other words, a particular novelty is the earliest instance of operation of a subsequently repetitive causal rule. Novelty as a metaphysical category refers to the fact that the number of repet-

itive causal rules in existence as a whole changes as a function of time. Novelty can only be identified retrospectively, because at the point of novelty, i.e. the instance of first operation of that causal rule, it is not possible to tell whether or not that rule will become repetitive or will cease to operate subsequently with any consistency (in which case it would be classifiable as chaotic).

Changes, and the causal rules that describe the connections between changes, are localized in particular regions of space and intervals of time. Any causal rule connecting particular changes is necessarily contextualized to a defined region of time and space; there is no such thing as a universally applicable, unified set of causal rules. With respect to a given region of space and interval of time, two categories of repetitive causal rules can be labeled; structure and function. (Chaotic causal rules are not relevant to either structure or function). Function is local repetition, structure is global repetition. "Local" in this sense means specific to the focal region of space and interval of time, "global" means occurring across the boundary of that region, i.e. relating that region to the wider context in which it occurs. Structure thus describes causal rules that connect changes outside the region to changes within it. Structural causal rules involve repetition that is larger in spatial scale and temporal duration than the frame of reference. Function describes causal rules that connect changes entirely within that frame of reference. Structural causal rules can affect functional causal rules, but not the other way around. The difference between structure and function is simply a way of differentiating between two different classes of causal rules, in terms of the distinct way in which these rules relate to the same region of time and space. A structure to function relation is a set of structural constraints imposed upon a set of functional rules. Such a structure-function relation can be labeled as a system. Two distinct systems may interact if they happen to be found in the same region of space and interval of time, and come into (structural and/or functional) contact. Such interactions happen for "random" reasons entirely unconnected with how either system works.

A number of possible outcomes can occur when two systems interact, only a subset of which give rise to novelty. A prerequisite for novelty is that an interaction occurs between systems that have similar function but different structures. Suppose that two systems satisfying this prerequisite come into contact and functionally interact. The functional similarity causes the repetition of the particular causal rules within the function of each system to become intertwined; i.e. their respective functional repetition falls into a synchrony. But the fact that the two systems are structurally different from each other means that, with respect to the function of each system, the structure of the other system is a chaotic input. This chaotic input causes particular changes to become associated with each other in time, in a way that they were not prior to the interaction between the systems. This association, i.e. between the "extra" structurally induced changes, and the original, repetitive functional changes, is itself chaotic (inconsistent in time). This is because there is no fixed relationship between the patterns of changes that are associated with each other by the causal rules of the structure of one system and the function of the other. So, we have the function of each single system simultaneously exhibiting both a repetitive, consistent relationship to the function of the other system, and a chaotic, inconsistent relationship to the structure of that same system. Novelty happens by a kind of "symbiosis" between systems exhibiting functional similarity and structural difference. The functional similarity between the two systems gives rise to a function-function interaction that responds to the chaotic/inconsistent aspect of the function-structure relationship. The (repetitive, consistent) function-function interaction causes some of the (initially chaotic, inconsistent) associations between structurally induced changes to become quasi-repetitive for brief periods of time (but not indefinitely). This is labeled transient repetition. This transient repetition of the structurally induced changes triggers the emergence of a structure-function relation at the level of both systems combined. This new structure-function relation is not reducible to either of the initial, starting systems. Some of the initially chaotic causal rules (i.e particular changes that were associated with each other inconsistently in time, as a result of the fact that the starting systems were structurally unrelated) become repetitive, as a result of the new structure-function relation that emerges. Turning a chaotic causal rule into a repetitive causal rule, such that this causal rule is contextualized to a particular region of time and space, effectively constitutes the origin of a new system. The structure of the new system is simply the sum of the structure of both initial systems. But the function constitutes the causal rules present in the functional interaction between the two starting systems, aswell as the initially chaotic, structurally induced "extra" causal rules that became fully repetitive when the new structure-function relation initially arose. Repetitive causal rules have, in effect, come into existence from nothing, such that changes have become associated with each other in a qualitatively new system.

This entire argument is thus simply a way of classifying causal rules in terms of when and where they occur in time. The number of repetitive causal rules in existence changes via discontinuous jumps. These changes relate to situations involving a combination of functional similarity and structural difference. The simple fact that systems arise from a contextualization of cause and effect, combined with fact that systems are spatially and temporally localized and may come into contact, creates an infinite potential for novelty. Potential novelty results simply from the transient associations between particular changes which occurs when systems come into contact "for no reason." But this transient novelty is indistinguishable from random, non-causal associations, due to its inconsistency in time. True novelty is the jump to consistent, repetitive operation of these initially chaotic, transiently repetitive causal rules. The essential prerequisite for such novelty is that the systems from whose interaction it results are sufficiently functionally similar to each other to assimilate the random structural inputs that occur. Two functionally dissimilar systems would lack this capacity for assimilation, i.e. lack the capacity to cause these random inputs to occur in a repetitive, ordered way. Novelty is more prevalent in contexts that force unrelated systems into the same time and space. The interaction between functionally similar but structurally different systems is rare, and a result so is novelty. When genuine novelty does occur it is also rare, and produces discrete, qualitative, often dramatic changes.

Expansion using a logical formalism

The essence of the argument is written in bold font at the start of each point in the following list. For the benefit of the non-mathematically inclined reader, the approximate meaning of each equation is written in words in the italicized text immediately before that equation. The indentation corresponds to the importance of the point, the most central points in the argument being indented to the left, more peripheral points to the right.

1. **Meaningful statements require the starting assumption of a foundational separation between "rules" and the "things" that are connected by such rules. It is unavoidable but acceptable that "rules" and "things" are defined tautologically relative to each other (i.e. rules are those relationships that relate things, and things are those entities that are related to each other by rules). Such a separation is a reasonable basis from which to develop a general perspective - one can speak in terms of rules and the things that they relate in a generic way by simply defining rules and things relative to each other, without the need for any external, foundational concept.** This is not the same thing as saying that this sort of distinction is more "real" than a system of thought based on something holistic, interactive, or irreducible. It is just that such holistic/irreducible systems of thought are less conducive to making clear meaningful statements. Potential examples of the unavoidable nature of this kind of linguistic separation include:
   (a) The tension between identity and change in formulating most philosophical problems[1]. Identity is necessary in order for us to be able to talk about something tangible, and change, or difference, is necessary for there to be something to talk about. Change in the most general sense can be though of as some sort of "time-ness" property exhibited by something with identity.
   (b) Category theory, the attempt at maximum generality in mathematically definable relationships, necessarily retains a separation between "arrows" and "objects", the former describing relationships between the latter[2]. Category theory refers to different mathematically defined relationships that have validity in different situations, and the idea behind category theory is to derive an "abstract algebra of functions" by means of which the mathematical relationships valid in one frame of reference can be related to those that are valid in another. Despite this aim for maximum generality, the mathematicians have found it necessary to hold on to a basic separation between things (objects) and the rules by means of which things are related (arrows).
   (c) In set theory, the axiom of separation, which states that the class of all sets is not itself a set[3]. The class of all sets is some sort of

context that allows a grouping of things into sets, and a definition of relationships between different sets. The class of all sets is a collection of rules, sets are the things that are subject to these rules. The context is necessarily a fundamentally different kind of entity from the things (sets) that are grounded within that context. A realization along these lines gave rise to Russell's paradox[4] which refers to a self-referential contradiction that arises when membership of a set can be assigned arbitrarily. Consider the set of all sets that are not members of themselves - is this set a member of itself? If it is, then it contains an element that is a member of itself, i.e. itself - and therefore violates the definitive criterion for elements belonging to that set. If is not a member of itself, then it excludes at least one set (itself) that is not a member of itself, and therefore violates the criteria by which it is defined; the inclusion of all sets that are not members of themselves. I.e *there exists set Q such that for any possible set x, x is not a member of itself implies that x is a member of Q:*

$$\exists Q \mid \forall x | x \notin x \Longrightarrow x \in Q, \tag{1}$$

The key point is that if sets can be members of themselves, then this statement should remain valid for all sets, including when $Q = x$, even when Q is defined as above. This straight away creates a contradiction. *Taking $Q = x$ implies Q (by the definition of Q in (1)) that Q is a member of Q and therefore is not a member of Q, thus taking $Q = x$ contradicts the existence of Q as defined by (1):*

$$Q = x \Longrightarrow Q \in Q \Longrightarrow Q \notin Q \Longrightarrow (Q = x) \perp \exists Q \tag{2}$$

One way out of this kind of self-referential mess is simply to decide that sets are fundamentally different from the things that are contained within them. This was the basis of Bertrand Russell's theory of types[4]. Within this sort of perspective, the above contradiction fails to arise, because sets simply cannot be defined arbitrarily with reference to anything and everything, they must be defined meaningfully. And sets, when meaningfully defined, cannot necessarily be the same kind of thing as the elements contained within them, it depends on the details of the properties of such things. Often, if the means by which a set is defined allows that set to be a member of itself, then the set of criteria is not meaningful and is unlikely to be much use (e.g. the notion of the "set of all sets" lacks much meaning aside from illustration of the above sort of self-referential abstraction). The lesson learned is that a foundational separation between elements (things) and sets (rules with respect to which things are collected together) must often be simply assumed if a meaningful notion of the grouping of things together into sets is to be retained.

(d) Chinese philosopher Lao Tse: "*The Tao is that through which all things have come into being, hence the Tao is not a thing*"[5]. Again, the Tao is presumably some sort of context by which things came to be, in other words some sort of set of rules relating things to each other. The mathematicians note a parallel between "the class of all sets/sets" and "Tao/things" separation[6].

(e) Wittgenstein's *Tractatus*[7]: "*The world is the totality of facts, not of things.*" This perspective shifts the emphasis to (internal) relationships, as opposed to objects considered in isolation. To quote Bertrand Russell's summary in the forward that he wrote to this book "*In order that a certain sentence should assert a certain fact there must, however the language may be constructed, be something in common between the structure of the sentence and the structure of the fact.*" This, of course, is a subtle issue, but the key point is that, one way or another, Wittgenstein began by attempting to shift the emphasis to relationships between entities rather than just the entities themselves. Say that if function of x $f(x)$ is a meaningful statement, in a language, about a real relationship $g(y)$ in the world, then the way in which function f relates to variable x in the language must be "intrinsically similar" to the way the real relationship g relates to thing y in the world. The idea that both statements in language and states of affairs in the world have internal structures, requires that they have rules governing which structures are permissible and which are not. Without getting distracted by the details of such a discussion, a prerequisite for this being the case is that there is some sort of difference between the type of entity a "fact" is and the type of entity "the intrinsic structure of that fact is" and these concepts must on some level be tautologically related to each other.

2. **Changes are a ubiquitous part of our experience, therefore are a good candidate for the most foundational kind of things that can be related to each other by rules. A change can be defined as anything with the combination of both identity and temporality.** Identity in this sense means that we can be sufficiently confident that the individual thing "exists", i.e. that there is some-"thing" there to talk about. In order to say that this thing is undergoing change, we must speak about temporality. Temporality means that the thing in question exhibits a recognizable, tangible relationship to time as we experience it. So a change is something that is both a thing that we can identify as part of our experience, and something that has a relationship to time. Thus, crudely speaking, identity means "thing-ness" and temporality means "change-ness", and a change is anything that has both these properties. Say that for any particular change C_p, the relationship that that particular change C_p has to time t is $f_{C_p}(t)$. Say that by definition C_p is genuinely a change if it has both identity $id_{C_p}(C_p) = C_p$, and temporality $C_p = f_{C_p}(t)$. Then we can write *there exists the category of all change* $\overline{C}$, *such that for all particular changes* C_p, *the existence of that particular change implies the existence of both an identity function*

for that change, and a function relating that change to time. Both these criteria together imply that C_p is a member of the category $\overline{C}$, (i.e. is genuinely a change):

$$\exists \overline{C} | \forall C_p, \exists C_p \Longrightarrow (\exists id_{C_p} | id_{C_p}(C_p) = C_p) \& (\exists f_{C_p} | C_p = f_{C_p}(t))$$
$$\Longrightarrow C_p \in \overline{C} \tag{3}$$

3. **Connections exist between particular changes, these connections can be described by causal rules. Causal rules describe the nature of the connection in time between the particular changes to which they refer. Causal rules can be classified into two mutually exclusive categories, repetition and chaos. Repetitive causal rules describe connections between particular changes that are consistent over time, chaotic causal rules describe describe connections between particular changes that are inconsistent over time.** Consider two particular changes C_j and C_k, which may or may not be associated with each other at a given point in time. If absolutely every time we observe change C_j we observe change C_k straight afterward, we would say that change C_j "causes" the occurrence of change C_k. We might be able to formalize a rule $r_{j,k}$ for how the connection between these changes works; i.e.$C_j \overset{r_{C_{j,k}}}{\longrightarrow} C_k$ (change C_j causes change C_k through mechanisms described by rule $r_{C_{j,k}}$). If we never observed C_k without first observing C_j, then we would confidently think in terms of such causation. This sort of straightforward, common sense notion of causal association between changes is repetition. By contrast, suppose sometimes we observed an association between the two changes and sometimes we did not, such that it was, in some circumstances, possible to see the second change C_k without it having been being preceded by the first C_j. This state of affairs would be more ambiguous, we would probably would not speak in terms of causation, we might question whether any association existed between the two changes at all, and would therefore not bother formalizing a "rule." If, in this second circumstance, we did decide that we wanted to write down a "rule" describing the (temporally inconsistent/erratic) association between the two changes, we might use phrases like "coincidence", "chance", "anomaly", "one-off-event" etc. This sort of inconsistent association within time between particular causal rules is chaos. *There exists the category of all causal rules $\overline{R_C}$, which is the sum of the category repetition $\overline{R_R}$ and chaos $\overline{R_K}$:*

$$\exists \overline{R_C} = \overline{R_R} + \overline{R_K} \tag{4}$$

A repetitive causal rule can be formally defined as follows. *Let changes C_j, C_k, both be members of the category change $\overline{C}$. If across all time t,* *there exists rule $r_{C_{j,k}}$ describing how and why the occurrence of change C_j always precedes the occurrence of C_k, i.e. $C_{j(t)} \overset{r_{C_{j,k}}}{\longrightarrow} C_{k(t+1)}$, then this rule $r_{C_{j,k}}$ describes repetitive causation, and is therefore a member of the category repetition $\overline{R_R}$:*

$$(C_j, C_k \in \overline{C})\&(\forall t, \exists r_{C_{j,k}} | (C_{j(t)} \overset{r_{C_{j,k}}}{\longrightarrow} C_{k(t+1)}) \Longrightarrow (r_{C_{j,k}} \in \overline{R_R}) \quad (5)$$

A chaotic causal rule can be formally defined as follows. *If two changes C_j and C_k, are both members of the category change $\overline{C}$, and at one point in time t_1 there exists a rule $r_{C_{j,k}}$ describing how and why C_j precedes C_k, but that at another later time point t_2 this rule does not operate (i.e. it is not true that C_j precedes C_k), then the rule $r_{C_{j,k}}$ describes inconsistent, non-repetitive causation, and is a member of the category chaos $\overline{R_K}$:*

$$(C_j, C_k \in \overline{C}), \&(t = t_1, \exists r_{C_{j,k}} | (C_{j(t_1)} \overset{r_{C_{j,k}}}{\longrightarrow} C_{k(t_1+1)})$$

$$\&(t = t_2 > t_1, (\nexists r_{C_{j,k}} \equiv \neg(C_{j(t_2)} \overset{r_{C_{j,k}}}{\longrightarrow} C_{k(t_2+1)}))) \Longrightarrow (r_{C_{j,k}} \in \overline{R_K})$$

(6)

By these two definitions, repetition and chaos are mutually exclusive; no causal rule is ever simultaneously repetitive and chaotic, and the intersection between the two categories is empty. *For all $r_{C_{j,k}}$ within $\overline{R_C}$, if $r_{C_{j,k}}$ is a member of the category repetition $\overline{R_R}$ it is not a member of the category chaos $\overline{R_K}$, and the converse; the intersection between these two categories $\overline{R_R} \cap \overline{R_K}$ is empty (contains no causal rules).*

$$\forall r_{C_{j,k}} \in \overline{R_C}, (r_{C_{j,k}} \in \overline{R_R} \Longrightarrow \neg r_{C_{j,k}} \in \overline{R_K})$$

$$\&(r_{C_{j,k}} \in \overline{R_K} \Longrightarrow \neg r_{C_{j,k}} \in \overline{R_R})$$

$$\Longrightarrow \overline{R_R} \cap \overline{R_K} = \emptyset \quad (7)$$

Thus, to reiterate, causal rules are particular associations between changes (which are elements of experience with both identity and temporality), repetitive causal rules operate with consistency across all time, chaotic causal rules operate inconsistently over time (such that chaotic causal rules apply at some points in time but not others), and no causal rule is ever both repetitive and chaotic simultaneously. A few footnotes:

(a) Repetitive causal rules linking other repetitive causal rules to each other are themselves repetitive causal rules. I.e. if one causal rule always operates prior to the operation of a second causal rule, we can define a third causal rule describing the link between them, and this third rule is also a causal rule, which might again be subject to higher order causal rules, etc, etc. This leads to a progressive and potentially infinite exercise of network building, founded upon what are ultimately associations between changes; which is one way of describing what science does.

i. Say that r_{C_j} and r_{C_k} both exist and are repetitive causal rules (members of $\overline{R_R}$), e.g. r_{C_j} describes connections between changes C_{j1} and C_{j2}, i.e. $C_{j1} \overset{r_{C_j}}{\longrightarrow} C_{j2}$, and rule r_{C_k} connects changes C_{k1} and C_{k2}, i.e. $C_{k1} \overset{r_{C_k}}{\longrightarrow} C_{k2}$. Say that the operation of rule r_{C_j} repetitively and consistently occurs prior to the operation of r_{C_k}. Then there exists a third rule $r_{j,k}$ describing how and why the operation of r_{C_j} causes (is repetitively associated with) that of r_{C_k}. This third rule $r_{j,k}$ is itself a repetitive causal rule (member of $\overline{R_R}$), and is equivalent to the sequential occurrence of the changes described by rule r_{C_j} followed by those described by rule r_{C_k}. *Say that r_{C_j} is a member of $\overline{R_R}$ and describes the association between changes C_{j1} and C_{j2}. Say that the operation of rule r_{C_j} implies that another repetitive causal rule r_{C_k} (describing the association between changes C_{k1} and C_{k2}) will operate at the subsequent time. This implies that there exists rule $r_{j,k}$ describing how and why the operation of rule r_{C_j} causes the operation of r_{C_k}. This third rule is also a repetitive causal rule, and amounts to a description of the sequential association between the particular time dependent changes $C_{k2} \circ C_{k1} \circ C_{j2} \circ C_{j1}$ described by the first two rules.*

$$\forall r, \forall t | (\exists r_{C_j}, r_{C_j} \in \overline{R_R}) \& (\exists r_{C_k}, r_{C_k} \in \overline{R_R})$$

$$| ((C_{j1(t-2)} \overset{r_{C_j}}{\longrightarrow} C_{j2(t-1)})$$

$$\& (r_{C_j(t-1)} \Rightarrow (C_{k1(t)} \overset{r_{C_k}}{\longrightarrow} C_{k2(t+1)})))$$

$$\Longrightarrow \exists r_{j,k} | r_{C_j} \overset{r_{j,k}}{\longrightarrow} r_{C_k}, r_{j,k} \in \overline{R_R} \tag{8}$$

$$r_{j,k} = r_{C_k} \circ r_{C_j} = C_{k2} \circ C_{k1} \circ C_{j2} \circ C_{j1} \tag{9}$$

(b) If we like, we might wish to use the word "event" to describe a particular instance of the operation of a specific causal rule at a specific point in time, i.e. to distinguish a rule's operation at one specific time point to its operation in a general sense. *Use the label $\overline{E}$ to denote the category of all events, all particular time-specific instances of operation of given causal rules. Say there exists a causal rule $r_{j,k}$ linking changes C_j and C_k $C_j \overset{r_{j,k}}{\longrightarrow} C_k$, and that there also exists a specific time point t_p such that the causal rule $r_{j,k}$ operated at that time point $C_{j(t_p)} \overset{r_{C_{j,k}(t_p)}}{\longrightarrow} C_{k(t_p+1)}$. This time-specific instance of operation of the rule $r_{C_{j,k}(t_p)}$ is an event, a member of the category $\overline{E}$:*

$$\forall j, k | (\exists r_{C_{j,k}} \in \overline{R_C}) \&$$

$$(\exists t_p | C_{j(t_p)} \overset{r_{C_{j,k}(t_p)}}{\longrightarrow} C_{k(t_p+1)}) \Longrightarrow r_{C_{j,k}(t_p)} \in \overline{E} \tag{10}$$

In more simple language, events are just a list of things that have happened that involved describable causal rules. Because time as we experience it only goes forward, the category containing all the events that have ever happened can only ever increase over time. *For all t, the size $|E|$ of the category $\overline{E}$ at time point t is always less than or equal to the size of the same category at the subsequent point in time $t+1$:*

$$\forall t, |\overline{E}|_{(t)} \leq |\overline{E}|_{(t+1)} \tag{11}$$

This directional accumulation of events in time might relate to the distinction that Bergson[8] made between "duration" (in terms of experience of events), and time (viewed as a dimension with no fundamentally different properties to space, as in Newtonian physics). Note that the increasing accumulation of events in time is not the same thing as novelty or "complexity."

(c) There is no way to prove that a repetitive causal rule that operates at the present time will continue to operate into the future, e.g.

Hume's problem of induction[9]. Science and induction-based reasoning deals with situations in which it possible to define and describe repetition, and ultimately treat the existence of such repetition as a starting assumption. It is a matter of taste as to whether one regards chaos as a real and unavoidable property of the world, or as a category containing all the causal rules that have not yet been discovered. Chaos is the label used here for what are one-off transient occurrences, which are not repetitive and therefore not amenable to science or induction-based reasoning. By the above definition literally any association between changes could be classified as being linked with chaotic "rules." Such chaotic rules would range from the trite "change one happened to coincide with change two for no real reason", to the more interesting "such-and-such a connection was a non-reproducible example of conscious synchronicity." Although one off events of synchronicity are not amenable to science, the fact that they are inconsistent and non-repetitive does not *necessarily* mean that they are not real. A dogmatic skeptic might dispute the reality of chaotic causal rules. But this would be because such a skeptic's definition of "real" roughly corresponded to "operates in accordance with repetitive causal rules", rendering such skepticism circular. The investigation of inconsistent association between changes is connected to what Jung called synchronicity, or meaningful coincidence[10].

4. **Novelty is the natural phenomenon whereby repetitive causal rules come into existence from a non-repetitive state. Novelty is real; the repetitive causal rules in operation within existence change over time.** For any repetitive causal rule that we might care to focus on, we can usually identify or infer the necessity of a point in the past at which that repetitive causal rule first began to operate. This goes hand in hand with saying that prior to this point in time the rule in question was not repetitive, and either was chaotic/inconsistent or simply did not exist (it is impossible to know the difference with reliability). A novelty is the first point in time at which a particular repetitive rule begins to operate. Novelty as a wider category is the disposition that exists for repetitive causal rules to begin operating from a previous background state of non-repetition. In other words, novelty is the property of nature that causes the total number of repetitive causal rules in existence to increase over time. *There exists the non-empty category novelty* $\overline{N}$. *If for all* t *greater than or equal to some time point* t_N, *there exists a causal rule* $r_{j,k}$ *(describing how/why change* C_j *precedes change* C_k*), and for all* t *less than* t_N *it is not true that this rule operates, and that the changes it describes are associated with each other, then for* $t < t_N$, *rule* $r_{j,k}$ *either does not exist at all or is chaotic/inconsistent (member of category* $\overline{R_K}$*), and if at time point* $t = t_N$ *onwards (for all time points* $t > t_N$ *) rule* $r_{j,k}$ *operates repetitively (is a member of category* $\overline{R_R}$*),*

then the time-specific first instance of operation of the repetitive causal rule $r_{j,k(t_N)}$ *can be retrospectively identified as an instance of novelty* $r_{j,k(t_N)} \in \overline{N}$.

$$\exists \overline{N} \neq \emptyset \tag{12}$$

$$(\forall t \geq t_N, \exists r_{j,k} | C_{j(t)} \xrightarrow{r_{j,k}} C_{k(t+1)}) \& (\forall t < t_N, \neg(C_{j(t)} \xrightarrow{r_{j,k}} C_{k(t+1)}))$$

$$\Longrightarrow (\nexists r_{j,k(t<t_N)}) \vee (r_{j,k(t<t_N)} \in \overline{R_K}) \&$$

$$(r_{j,k(t=t_N)} \in \overline{N}) \& (r_{j,k(t>t_N)} \in \overline{R_R}) \tag{13}$$

Thus, a novelty is the first instance of operation of a subsequently repetitive causal rule. Because the category novelty as thus defined contains specific operation of causal rules at specific points in time (i.e. rather than containing these rules in general), the category novelty is contained within the category events; all novelties are events, but not all events are novelties. *The category* $\overline{N}$ *of all novelty (all particular first instances of operation of specific causal rules at specific points in time) is a subset of the category of events* $\overline{E}$ *(all particular instances of the operation of causal rules at particular points in time):*

$$\overline{N} \subset \overline{E} \tag{14}$$

A few conceptual points are relevant to novelty as thus defined:

(a) Novelty is the connection between chaos and repetition. The non-repetitive "background state" of a causal rule, prior to an instance of novelty, is, in practical/scientific terms indistinguishable from its non-existence. Only the novelty that gives rise to repetition is knowable, because only repetitive features of the observable world are knowable in any meaningful sense. Transient novelty (i.e. causal rules that operate for the first time at one particular time point, but do not subsequently become repetitive) cannot be ruled out or said to not exist, but nor can it be reliably known to exist. Knowable novelty refers to those rules that make the jump between chaos and repetition.

(b) All novelties are events, not rules or points in time. A rule can only ever be a novelty at the first point in time at which it operates, after which it either ceases to operate or becomes repetitive. Thus, any given novelty is a tie between a rule and a specific point in time at which the repetition described by that rule initiates - neither the rule nor the point in time are novel in themselves, the novelty is the connection between the two. This gives novelty a fundamentally contextual, contingent property.

(c) Novelty is fundamentally unpredictable, only identifiable retrospectively, and in some ways beyond the scope of the scientific method. At the first point of a rule's operation, the time point at which it is actually is a novelty, one cannot identify it as a causal rule (because doing so requires the rule's repetitive operation, which can only be identified later). Strictly speaking, novelty is therefore outside the scientific method (or anything else that rests on inductive extrapolation). However, sometimes one can identify a repetitive causal rule and "back track" to its necessarily having been a novelty at a previous point in time. Indeed, the fact that much of historical science does this is the backbone of the evidence for novelty being a real property of nature. But all the scientific method can do is take a repetitive causal rule and infer that it must have initiated as a novelty; novelty cannot be predicted before the fact (because such prediction relies on the causal rules in existence prior to the novelty, which, by definition, the novelty is not reducible to).

(d) Even if novelty is dismissed as an unfolding of repetition within some underlying cycle of recurrence, there still remains the question of how that unfolding works, meaning that novelty and creativity remain legitimate problems. To dismiss novelty as recurrence is simply to relabel it as repetition that we were previously unaware of, by virtue of an appeal to an underlying repetitive cycle which is by definition unobservable directly. Such a dismissal is neither parsimonious nor useful.

5. **Changes, and as a result the causal rules that connect them, can be localized in space as well as time.** For any given change, it is possible to identify a region of space in which that change occurs. Recall that a change is any aspect of experience with the combination of identity and temporality. The identity aspect of a change (for example "being an entity that has consistent physical properties", labeled as $id_{C_q}(C_q) = C_q$ for change C_q) is what can be localized in space. So $id_{C_q}(C_q) \subset \overline{S_q}$, meaning "the identity aspect of change C_q is a subset of region of space $\overline{S_q}$", simply means that this change happens in this region of space $\overline{S_q}$. *For all particular changes C_q (within the category of all change $\overline{C}$), there exists a region of space S_q, such that the identity property $id_{C_q}(C_q)$ of any particular change C_q is a subset of that region of space:*

$$\forall C_q \in \overline{C}, \exists \overline{S}_q | id_{C_q}(C_q) \subset \overline{S_q} \tag{15}$$

In other words a change is not a change unless it happens somewhere specific. (Because if it did not happen somewhere specific it would not exhibit sufficient identity to count as a change). Consequently, when a rule describes a connection between changes that are spatially localized, this rule is also spatially localized. *If change C_n is contained in region of space S_1, $id_{C_n}(C_n) \subset \overline{S_1}$ and so is change C_m, $id_{C_m}(C_m) \subset \overline{S_1}$, and if there exists a causal rule $r_{C_{n,m}}$ linking these* *changes (describing how and why change C_n occurs first and leads to change C_m, $C_{n(t)} \xrightarrow{r_{C_{n,m}}} C_{m(t+1)}$), then this rule $r_{C_{n,m}}$ is also contained within region of space $\overline{S_1}$, $r_{C_{n,m}} \subset \overline{S_1}$:*

$$(id_{C_n}(C_n) \subset \overline{S_1})\&(id_{C_m}(C_m) \subset \overline{S_1})\&$$

$$(\exists r_{C_{n,m}} | (C_{n(t)} \xrightarrow{r_{C_{n,m}}} C_{m(t+1)}) \Longrightarrow r_{C_{n,m}} \subset \overline{S_1} \quad (16)$$

This kind of spatial contextualization of causal rules is the crux of the distinction between structure and function.

6. **A system is a collection of repetitive, interconnected causal rules, and the spatial/temporal frame of reference within which those rules operate. Function is local repetition and structure is global repetition, where "local" and "global" are defined with respect to the spatial and temporal boundaries of a given system, and a particular system is defined (circularly) as a localized connection between a particular structure and function.** Say the region of space S_1 marks out the spatial boundaries of something we can meaningfully classify as a particular system σ_1, and say that this system persists over timescales of the order of a particular time interval $\triangle t_1$. The first way in which we can distinguish structure and function is in terms of the spatial frame of reference in which particular changes are connected, the second is with respect to the timescale over which such rules operate.
   (a) **Function refers to repetitive causal rules operating inside a system, whereas structure refers to repetitive causal rules connecting the system to the outside.** A given causal rule is part of the function of system if it connects two changes that are both localized inside the spatial boundary of the system. *Say that there exists repetitive causal rule $r_{C_{j,k}}$ describing the connection between changes C_j and C_k, and both these changes are localized within region of space $\overline{S_1}$ $(C_j \subset \overline{S_1})\&(C_k \subset \overline{S_1})$. This causal rule is therefore part of the function $\overline{F_{\sigma_1}}$ of the system σ_1 within that region of space $\overline{S_1}$. This amounts to saying that the function of the system is congruent with that region of space $\overline{F_{\sigma_1}} \cong \overline{S_1}$, i.e. that the function of the system can be defined relative to the boundaries of that region of space.*

$$(\exists r_{C_{j,k}} | (C_{j(t)} \xrightarrow{r_{C_{j,k}}} C_{k(t+1)})\&(C_j \subset \overline{S_1})\&(C_k \subset \overline{S_1})$$

$$\Longrightarrow r_{C_{j.k}} \in \overline{F_{\sigma_1}} \Longrightarrow (\overline{F_{\sigma_1}} \cong \overline{S_1}) \quad (17)$$

By contrast, a given causal rule is part of the structure of system σ_1 if it connects a change from outside that system (outside region of space $\overline{S_1}$, i.e. within $\overline{S_1}^C$, the complement of category $\overline{S_1}$) to a change inside the system (inside $\overline{S_1}$). *Say there exists causal rule $r_{C_{j.k}}$ describing the connection between changes C_j and C_k, and one of these changes is localized within a region of space $\overline{S_1}$ but the other is not ($C_j \subset \overline{S_1}^C \& C_k \subset \overline{S_1}$) or ($C_j \subset \overline{S_1} \& C_k \subset \overline{S_1}^C$). This implies that this causal rule is part of the structure $\overline{St_{\sigma_1}}$ of system σ_1. This, similarly amounts to saying that that structure $\overline{St_{\sigma_1}}$ is contained with that region of space, is compatible, or "congruent" with it $\overline{St_{\sigma_1}} \cong \overline{S_1}$:*

$$(\exists r_{C_{j,k}} | (C_{j(t)} \overset{r_{C_{j,k}}}{\longrightarrow} C_{k(t+1)})$$

$$\&(C_j \subset \overline{S_1}^C \& C_k \subset \overline{S_1}) \vee (C_j \subset \overline{S_1} \& C_k \subset \overline{S_1}^C)$$

$$\Longrightarrow r_{C_{j.k}} \in \overline{St_{\sigma_1}} \Longrightarrow \overline{St_{\sigma_1}} \cong \overline{S_1} \quad (18)$$

So the key aspect of structure is the establishment of causal connection between a system and its exterior. This can be in both directions structure can refer to a change outside the system causing a change inside it, or vice versa.

(b) **Functional causal rules operate intermittently over the timescale within which it is possible to define a system; i.e. the period of repetition of functional causal rules is comparable to the time interval over which the system persists. Structural causal rules are constant over this timescale; i.e. the period of repetition of structural causal rules is shorter than this time interval.** For instance, with respect to the characteristic time interval over which a system can be said to tangibly exist, functional causal rules operate repetitively but intermittently, such that they switch on and off in relation to each other, potentially in an infinitely complex, network-like way. This, of course, is what is meant by saying that a system has a function. By contrast, structural causal rules operate in a constant, "background" manner, such that at every time point within the time interval over which the system operates, all structural causal rules operate. The frequency of the repetition of structural causal rules is higher than that of functional rules (but note that both structural and functional causal rules are repetitive rather than chaotic, the difference between the two classes of rules refers to the period of repetition, relative to the time interval that characterizes the system). The characteristic time interval of the system is defined such that any given functional causal rule operates at least once during that time interval. *Causal rule* $r_{C_{j,k}}$ *is a member of the category function* $\overline{F_{\sigma_1}}$ *of system* σ_1*, which exists over characteristic time interval* $\overline{\Delta t_1}$*. This implies that there exists the set of time points* $\{t_{off}\}$ *such that for all particular time points* t_m *within* $\{t_{off}\}$*, at that time point it is not true that causal rule* $r_{C_{j,k}}$ *is operating. There also exists the set of time points* $\{t_{on}\}$ *such that for all particular time points* t_m *within* $\{t_{on}\}$ *it is true that causal rule* $r_{C_{j,k}}$ *is operating. Obviously, for any time point within the characteristic time interval of the system* $\forall t_m \in \Delta t_1$*, if this time point is a member of the set* $\{t_{off}\}$ *this implies that it is not a member of the set* $\{t_{on}\}$*, and the other way around, and all time points within the characteristic time interval* $\forall t_m \in \overline{\Delta t_1}$ *are either a member of* $\{t_{off}\}$ *or of* $\{t_{on}\}$*. When all these things are true, we say that the function is compatible or congruent with that interval of time* $\overline{F_{\sigma_1}} \cong \overline{\Delta t_1}$*:*

$$\forall r_{C_{j,k}} \in \overline{F_{\sigma_1}}, (\exists[\{t_{off}\}|\forall t_m \in \{t_{off}\} \Longrightarrow \neg(C_{j(t_m)} \overset{r_{C_{j,k}}}{\longrightarrow} C_{k(t_m+1)})$$

$$\&(\exists\{t_{on}\}|\forall t_m \in \{t_{on}\} \Longrightarrow (C_{j(t_m)} \overset{r_{C_{j,k}}}{\longrightarrow} C_{k(t_m+1)})$$

$$\&\forall t_m \in \overline{\Delta t_1}, t_m \in \{t_{off}\} \Longrightarrow$$
$$t_m \notin \{t_{on}\}\&t_m \in \{t_{on}\} \Longrightarrow t_m \notin \{t_{off}\}$$

$$\&\forall t_m \in \overline{\Delta t_1}, t_m \in \{t_{off}\} \vee t_m \in \{t_{on}\} \Longrightarrow \overline{F_{\sigma_1}} \cong \overline{\Delta t_1} \qquad (19)$$

By contrast, a causal rule that is part of the structure of a system is always "on" at every time point within that system's characteristic time interval. *Causal rule* $r_{C_{j,k}}$ *(i.e. connecting changes* C_j *and* C_k *as labeled above) is a member of the category structure* $\overline{St_{\sigma_1}}$ *of system* σ_1*. This implies that for all time points within the system's characteristic time interval* $\forall t_m \in \overline{\Delta t_1}$*, this rule operates at that time point. This implies that the structure* $\overline{St_{\sigma_1}}$ *of system* σ_1 *is congruent with interval of time* $\overline{\Delta t_1}$*:*

$$r_{C_{j,k}} \in \overline{St_{\sigma_1}} \Longrightarrow \forall t_m \in \overline{\Delta t_1}, C_{j(t_m)} \overset{r_{C_{j,k}}}{\longrightarrow} C_{k(t_m+1)}$$

$$\Longrightarrow \overline{St_{\sigma_1}} \cong \overline{\triangle t_1} \tag{20}$$

Thus, for the specific interval of time relevant to the definition of a system, functional causal rules can be on or off, structural causal rules are constantly on. Time is thus a more integral part of function than it is of structure. Structure is a background constant, structural repetition constitutes a reference frame that grounds the system. Function is cyclical, functional repetition is what the system does in terms of interaction with the exterior. *Saying that the structure and the function of a system are both congruent with an interval of time and a region of space is the same thing as saying that that whole system is congruent with that region of space and time:*

$$(St_{\sigma_1} \cong \overline{\triangle t_1})\&(\overline{F_{\sigma_1}} \cong \overline{\triangle t_1})\&(\overline{F_{\sigma_1}} \cong \overline{S_1})\&(\overline{St_{\sigma_1}} \cong \overline{S_1})$$

$$\equiv \sigma_1 \cong (\overline{S_1}, \overline{\triangle t_1}) \tag{21}$$

(c) **A system can exist for multiple iterations of its characteristic time interval, and functional repetition is self referential during these iterations**. The system's characteristic time interval $\overline{\triangle t_1}$ thus represents one "cycle" of the system, in the sense that any given functional rule must, by definition, operate at least once within this time interval. Any particular system exists for multiple iterations of its characteristic time interval. Particular sequences of the operation of functional causal rules (i.e. in different relative orders) can be labeled as an aggregate "state" of that system's function. The function of a given system relates to itself over time through such aggregate states, such networks of operation of functional causal rules. *Use the letter α to denote a sequence of operation of particular functional causal rules within system σ_1, so that $\alpha_j = \prod_{k=1}^{j} r_{C_{k-1,k}}$ is a long sequence of causal rule operations, the combined result of the operation of causal rule $r_{C_{0,1}}$ then $r_{C_{1,2}}$ then $r_{C_{2,3}}$ etc, all the way up to change j $r_{C_{j-1,j}}$ for some particular sequence of j changes, in which all the causal rules connecting each change within the sequence are part of the function $\overline{F_{\sigma_1}}$ of system σ_1, $\forall r_{C_{n,m}} \in \alpha_j, r_{C_{n,m}} \in \overline{F_{\sigma_1}}$:*

$$\alpha_j = \prod_{k=1}^{j} r_{C_{k-1,k}} = r_{C_{j-1,j}} \ldots\ldots r_{C_{2,3}} \circ r_{C_{1,2}} \circ r_{C_{0,1}}$$

$$|\forall r_{C_{n,m}} \in \alpha_j, r_{C_{n,m}} \in \overline{F_{\sigma_1}} \tag{22}$$

Within the function of any system there is a category containing all the specific sequences of changes that the causal rules within that function refer to and connect to each other. This category of all possible particular sequences of causal rule operations within the function of the system, constitutes the system's aggregate functional state $\overline{\alpha_{\sigma_1}}$. It is through this aggregate functional state that the system's function relates to itself, responds to the world with an appropriate level of variety and flexibility, etc. The aggregate functional state $\overline{\alpha_{\sigma_1}}$ is thus a label for what the system actually does over time, relative to itself and its surroundings. *There exists aggregate functional state $\overline{\alpha_{\sigma_1}}$ for system σ_1, such that for all specific sequences α_j of repetitive causal rule operations within the function $\overline{F_{\sigma_1}}$ (of system σ_1), any such sequence is a member of that aggregate functional state $\alpha_j \in \overline{\alpha_{\sigma_1}}$:*

$$\exists\overline{\alpha_{\sigma_1}}|\forall\alpha_j \in \overline{\alpha_{\sigma_1}} \tag{23}$$

The aggregate functional state of a system is, in the language of category theory an "endomap" (a rule relating something to itself) for that system; the functional repetition of that system can be said to refer to itself through the aggregate functional state. *For all/any system σ_j, the existence of an aggregate functional state for that system $\exists\overline{\alpha_{\sigma_j}}$ implies that the function of the system $\overline{F_{\sigma_1}}$ relates to itself through that aggregate functional state $\overline{F_{\sigma_j}}^{\circlearrowleft\overline{\alpha_{\sigma_j}}}$:*

$$\forall\sigma_j, \exists\overline{\alpha_{\sigma_j}} \Longrightarrow \overline{F_{\sigma_j}}^{\circlearrowleft\overline{\alpha_{\sigma_j}}} \tag{24}$$

This self-referential characteristic, and the potential complexity that it produces, is the key distinction between structure and function. Function is self-referential and temporally flexible, structure is a background reference frame.

(d) **The existence of both a function with an aggregate functional state, and a structure, is equivalent to the existence of a system.** We can define a system with respect to a characteristic region of space and a characteristic interval of time. The function and structure of that system are just categories containing particular collections of causal rules that operate in that region of space and time. Function, localized repetition, is what a system does. Structure, globalized repetition, refers to constant causal rules imposed on a system as a consequence of the material that system is built from and the context in which it occurs. *With respect to system σ_1 in region of space $\overline{S_1}$ and time $\overline{\Delta t_1}$, saying that there exists the function of this system $\exists\overline{F_{\sigma_1}} \neq \emptyset$ (it is not equal to the empty set, i.e. it contains at least some causal rules), and there exists aggregate functional state $\overline{\alpha_{\sigma_1}}$ and structure $\overline{St_{\sigma_1}}$, is equivalent to saying that there exists the system $\sigma_1$itself:*

$$(\exists\overline{F_{\sigma_1}} \neq \emptyset)\&(\exists\overline{\alpha_{\sigma_1}} \neq \emptyset)\&(\exists\overline{St_{\sigma_1}} \neq \emptyset) \equiv \exists\sigma_1 \tag{25}$$

This is true not just for the specific system σ_1 but for any system in general σ_j:

$$(\exists\overline{F_{\sigma_j}} \neq \emptyset)\&(\exists\overline{St_{\sigma_j}} \neq \emptyset)\&(\exists\overline{\alpha_{\sigma_j}} \neq \emptyset) \equiv \exists\sigma_j, \forall j \qquad (26)$$

7. **Structure affects function.** With respect to the spatial/temporal boundary of a system, structural rules are imposed on the system from the outside, whereas functional rules are internal. Structure limits function, and grounds the events that happen within systems in "real stuff" (i.e. the material constituents of that system, biasing the range of functional events that are "allowed" within that system). *For all systems, there exists a structure to function relation* $r_{(\overline{St \Longrightarrow F})_{\sigma_j}}$, *such that structure* $\overline{St_{\sigma_1}}$ *affects the function of that system* $\overline{F_{\sigma_j}}$ *and its self-referential properties* $\overline{F_{\sigma_j}}^{\circlearrowleft\overline{\alpha_{\sigma_j}}}$ *through this relation:*

$$\forall\sigma_j, \exists r_{(\overline{St \Longrightarrow F})_{\sigma_j}} | \overline{St_{\sigma_j}} \overset{r_{(\overline{St \Longrightarrow F})_{\sigma_j}}}{\longrightarrow} \overline{F_{\sigma_j}}^{\circlearrowleft\overline{\alpha_{\sigma_j}}} \qquad (27)$$

I use the word "relation" rather than "rule" here because the structure function relation $r_{(\overline{St \Longrightarrow F})_{\sigma_j}}$ is not a causal rule as defined above, but rather a constraint on the operation of other rules. One might say that the functional aggregate state $\overline{\alpha_{\sigma_j}}$ of an actual real system σ_j, is a subset of a larger "potential" functional aggregate state for that system. Not all of this potential functional aggregate state is realized in the real system, due to constraints imposed upon function by structure. The potential state might be consistent with how the system functions (e.g. it might have causal rules that would tie in to those contained in the system's function in an ordered way, might be "useful" to that system etc), but aspects of it are constrained so as to not exist in the real system, and these constraints are the structure to function relation. *There exists (for all systems* $\forall\sigma_j$*) the potential aggregate functional state* $\overline{\alpha_{\sigma_j,POTENTIAL}}$ *of the system* σ_j*, of which the actual, realized aggregate functional state* $\overline{\alpha_{\sigma_j}}$ *of system* σ_j *is a subset* $\overline{\alpha_{\sigma_j}} \subset \overline{\alpha_{\sigma_j,POTENTIAL}}$. *The structure to function relation of system* σ_j $r_{(\overline{St \Longrightarrow F})_{\sigma_j}}$ *negates the existence of the intersection between the potential aggregate functional state and the complement the actual functional state* $\overline{\alpha_{\sigma_j,POTENTIAL}} \cap \overline{\alpha_{\sigma_j}}^C$. *This means that any causal rule* $r_{n,m}$ *within this intersection* $r_{n,m} \in \overline{\alpha_{\sigma_j,POTENTIAL}} \cap \overline{\alpha_{\sigma_j}}^C$, *is not a member of the actual function of the system* $r_{n,m} \notin \overline{F_{\sigma_j}}$, *(for any/all such causal rule* $\forall r_{n,m}$*). This is equivalent to saying that potential aggregate functional state* $\overline{\alpha_{\sigma_j,POTENTIAL}}$ *is "converted" to the actual aggregate functional state* $\overline{\alpha_{\sigma_j}}$ *through the structure to function relation* $r_{(\overline{St \Longrightarrow F})_{\sigma_j}}$:

$$\exists\overline{\alpha_{\sigma_j,POTENTIAL}} | (\overline{\alpha_{\sigma_j}} \subset \overline{\alpha_{\sigma_j,POTENTIAL}})$$

$$\&(|\overline{\alpha_{\sigma_j,POTENTIAL}}|\gg|\overline{\alpha_{\sigma_j}}|),\forall\sigma_j$$

$$\&(r_{(\overline{St\Longrightarrow F})_{\sigma_j}}\neg\overline{\alpha_{\sigma_j,POTENTIAL}}\cap\overline{\alpha_{\sigma_j}}^C|r_{n,m}\in\overline{\alpha_{\sigma_j,POTENTIAL}}\cap\overline{\alpha_{\sigma_j}}^C$$

$$\Longrightarrow r_{n,m}\notin\overline{F_{\sigma_j}},\forall r_{n,m}) \tag{28}$$

$$\equiv \overline{\alpha_{\sigma_j,POTENTIAL}} \xrightarrow{r_{(\overline{St\Longrightarrow F})_{\sigma_j}}} \overline{\alpha_{\sigma_j}} \tag{29}$$

Thus structure modifies function from what it could be, to what it actually is, through the structure function relation, by directly preventing functional possibilities. More simply, structural constraints restrict what is functionally possible to what is both functionally and structurally possible.

8. **Function cannot affect structure.** The relationship between the two sets of rules is asymmetric; structure constrains function, not the other way around. Functional rules modify the activity of other functional rules, but functional rules do not modify structural rules, they are subject to them. *There does not exist a rule* $r_{(\overline{F\Longrightarrow St})_{\sigma_j}}$ *by which the particular function of a system* F_S *modifies the fundamental structure* St_S *of the system*:

$$\nexists r_{(\overline{F\Longrightarrow St})_{\sigma_j}} \mid \overline{F_{\sigma_j}} \xrightarrow{r_{(\overline{F\Longrightarrow St})_{\sigma_j}}} \overline{St_{\sigma_j}} \tag{30}$$

This is the key difference between structure and function. It is results from the fact that functional causal rules are spatially contextual, i.e. grounded in a particular spatial/temporal context that they cannot step outside or modify - and structural causal rules describe/explain the nature of this context.

9. **One cannot have a system that has no constraints upon it. There is therefore no "universal system" across all space and time, because it would be impossible for such a system to have structure; all systems function locally.** Imagine a system with a "border" that encompassed all of space. Saying that such a system exists amounts to saying that it has structure, function, and a structure-function relation. For this to be the case there would have to be an "outside" of this system that imposed structural constraints upon it. This would require either that there was something outside the totality of space, or that the system had no structural constraints on its function, and therefore could contain any causal rule. If there is something outside the entirety of space, the concept of space becomes meaningless (or at least so far removed from a common sense notion of space as to warrant our using a completely different word). If the real universe

could do anything, if there were no restrictions on causal rules within it, we would notice (and indeed advocates of the anthropic principle might argue that we would not exist at all). All systems, and the operation of the causal rules that constitute them, are contingent upon a localized, defined region of space and a characteristic timescale. *There exists region of space $\overline{S_U}$, such that for all other definable regions of space $\overline{S_j}$ for all j, this definable space is a subset of $\overline{S_U}$ (i.e. such that $\overline{S_U}$ constitutes all space). This implies that there does does not exist a structure $\overline{St_{\sigma_U}}$ corresponding to a system σ_U with spatial boundaries defined by the totality of space $\overline{S_U}$. This in turn implies that there does not exist any such system σ_U encompassing the entirety of space, because it would be impossible for such a system to have a structure $\overline{St_{\sigma_U}}$, structure to function relation $r_{(\overline{St \Longrightarrow F})_{\sigma_U}}$, function $\overline{F_{\sigma_U}}$ with aggregate functional state $\overline{\alpha_{\sigma_U}}$, etc):*

$$\exists \overline{S_U} \mid \forall j, \overline{S_j} \subset \overline{S_U} \Longrightarrow \nexists \overline{St_{\sigma_U}} \Longrightarrow \nexists \sigma_U \mid \overline{St_{\sigma_U}} \overset{r_{(\overline{St \Longrightarrow F})_{\sigma_U}}}{\longrightarrow} \overline{F_{\sigma_U}}^{\circlearrowright \overline{\alpha_{\sigma_U}}} \tag{31}$$

In other words, spatial existence as a whole cannot be considered a system in any meaningful sense. A few footnotes:

(a) Whether a causal rule is structural or functional depends on the relationship that the rule in question has to a particular region of time and space. The same causal rule might be structural or functional depending on which system one is talking about. Different systems can share rules (i.e. be "copies" of each other) but no two systems can share exactly the same spatial and temporal boundaries (or the distinction between matter and space itself becomes meaningless). If two systems are copies of each other occupying different regions of space and time, there still arises the question as to when the earliest instance of the form of that copy came into existence (in other words saying that copies of different systems exist does not negate the question of novelty).

(b) Space, in the labeling system developed here, is a frame of reference that constrains which particular changes and the causal rules connecting them can and cannot happen, whereas time is the medium inside which events actually do happen. So the assumption is that space (through its connection to matter and material causation) is more deeply connected to how constraints upon causation occur than is time, i.e. that if the past influences the present, it does so through matter and space. If we relaxed the definition of structure so as to include "any constraint on what a system can do", without being specific to matter and space, we might be able to include hypothetical non-material models of causation, such as Sheldrake's model of formative causation[11] without spatial/material contact.

In this case it might be possible to imagine a system encompassing all space. However one would still need constraints on "what gets repeated" over time in such a model. Even in such hypothetical scenarios, the key point is that the concept of system becomes meaningless without constraints on the operation of causal rules.

(c) With respect to the real universe, a valid alternative argument might be to take the view that the entirety of the universe is indeed a single system, which does have constraints on it (constraints that would be classified as "structural" in the above labeling scheme), but that these constraints are unknowable. Bohm discusses an idea like this in the context of a distinction between the "implicate" and "explicate" order[12]. Crudely speaking, the explicate order is the manifested universe, which is an unfolding of an implicate (not directly observable, essentially unknowable) order. If one takes this view a system encompassing all space (and time) can be envisaged, but the constraints upon such a "system" cannot be known, so labeling it as such becomes meaningless. There is a loose parallel here to the sentiments expressed by Feyerabend: "*We have to realize that a unified theory of the physical world simply does not exist. We have theories that work in restricted regions, we have purely formal attempts to condense them into a single formula, we have lots of unfounded claims (such as the claim that all of chemistry can be reduced to physics), phenomena that do not fit into the accepted framework are suppressed. In physics, which many scientists regard as the one really basic science, we have now at least three different point of view (relativity, dealing with the very large, quantum theory for an intermediate domain and various particle models for the very small) without a promise of conceptual (and not only formal) unification.*"[13] One cannot have a unified system without constraints on that system, and one cannot have constraints outside the entirety of existence. Therefore the entirety of existence cannot be a single system in any meaningful sense.

10. **Novelty happens at the origin of systems; new causal rules operate for the first time at the point in time at which a connection between structure and function originates. This is why novelty is relatively rare, and tends to occur in a dramatic, step-like fashion.** Causal rules do not simply begin existing and then continue to exist, the origin of causal rules goes hand in hand with the origin of entirely new systems. Causal rules relate changes, which are grounded in something with identity (most obviously matter within space) and therefore subject to constraints. So the crux of the question of novelty is *"where do causal rules come from?"*, which needs to be answered in conjunction with the related question *"where do constraints on causal rules come from?"* The constraints that are imposed on causal rules, in terms of which changes are associated with each other and which are not, are what we have defined as structure, and a connection

between structure and function is what we have defined as a system. So the question of novelty is essentially the same thing as the question *"where do systems come from?"*

11. **Novelty results from the interaction between systems with similar function and different structure. The occurrence of a specific sequence of events during such an interaction is necessary and sufficient for novelty to occur.** If two systems that function in similar ways happen to interact in time and space, a relationship is established such that the functional repetition of each system becomes non-random with respect to that of the other - i.e. the functional repetition of each system "falls into synchrony" with that of the other. If the same two systems are also structured differently, each system is continuously exposed to what is, in effect, a random input (i.e. the uncorrelated structure of the other system). The function of each system responds to this input, such that transient associations between changes occur that are not part of that system's function, but are a sort of chaotic "emergency response" to the randomness of the external, unrelated structure. This transient repetition, of what are, in effect, quasi causal rules, leads to the emergence of a between-system functional relationship containing causal rules that were not present in either of the initial systems. These new causal rules initially occurred transiently, as a result of random changes being associated with each other by the impact of the structure of one system upon the function of the other. Assimilation of these initially chaotic causal rules, such that they subsequently become fully repetitive, goes hand in hand with the establishment of a structure to function relationship, at the level of both initial systems. The function of the new system contains causal rules that did not exist prior to the onset of the interaction between the two systems; it contains novelty, as a result of a kind of symbiosis between the two initial systems. The structure of the new system is the sum of that of the initial two systems. Both functional similarity and structural difference are prerequisites for this sequence of events to happen. (If the two systems are not functionally similar, the interaction is just structural, and any interactive events are not really localized to the systems. And if the two systems are not structurally different, then there is no random external influence for either system to respond to, therefore no potential for unprecedented association between changes). In other words, random structural input associates previously non-associated changes, functional similarity assimilates some of these associations so as to make them permanent, this assimilation entails the emergence of a new structure-function relation. Consider two systems σ_1 and σ_2, each with all the properties discussed above (structure, function, aggregate functional state, spatial boundaries, characteristic timescale, material basis, etc). Assume the two systems are structurally unrelated and functionally related to each other. Then the necessary and sufficient sequence of events for the interaction

between these two systems to give rise to novelty is as follows:

(a) **The two systems interact in space and time (for "random" reasons unconnected with either system).** The interaction is two-way; both systems influence each other. *There exists interaction $\overline{I_{\sigma_1 \Leftrightarrow \sigma_2}}$ such that via this interaction, system one σ_1 influences system two σ_2, and system two influences system one:*

$$\exists \overline{I_{\sigma_1 \Leftrightarrow \sigma_2}} \mid \sigma_1 \quad \overset{\overline{I_{\sigma_1 \Leftrightarrow \sigma_2}}}{\underset{\overline{I_{\sigma_1 \Leftrightarrow \sigma_2}}}{\overset{\longrightarrow}{\longleftarrow}}} \quad \sigma_2 \tag{32}$$

Obviously, for two systems to interact, they must overlap in space and time. Recall that above we defined a system as congruent with a particular region of space if the "inside" and "outside" of that region distinguishes "local" and "global" in the structure and function of that system. Similarly we defined a systems as congruent with an interval of time if the characteristic timescale over which the functional repetition of that system happens is comparable to that interval of time. In such a case we would say that the whole system is congruent with that region of space and interval of time, labeled as $\sigma_a \cong (\overline{S_b}, \overline{\triangle t_c})$, i.e. read "system a is congruent with region of space $\overline{S_b}$ and interval of time $\overline{\triangle t_c}$." Consider region of space $\overline{S_j}$ and interval of time $\overline{\triangle t_j}$, define the set $\overline{\sigma_{S_j, \triangle t_j}}$ of all systems that are congruent with this region of space and interval of time. Crudely speaking this set is just the collection of all the systems that are "in" that region of space and time. *There exists set $\overline{\sigma_{S_j, \triangle t_j}}$, such that a particular system σ_k (for all k) being a member set $\overline{\sigma_{S_j, \triangle t_j}}$ implies that this system is congruent with space/time interval $(\overline{S_j}, \overline{\triangle t_j})$:*

$$\exists \overline{\sigma_{S_j, \triangle t_j}} \mid \forall k. \, \sigma_k \in \overline{\sigma_{S_j, \triangle t_j}} \implies \sigma_k \cong (\overline{S_j}, \overline{\triangle t_j}) \tag{33}$$

If two systems are interacting this implies, by necessity, that they are both congruent with the same region of space and interval of time. *For all N, M such that there exists interaction $\overline{I_{\sigma_N \Leftrightarrow \sigma_M}}$ between systems σ_N and σ_M, This implies that there exists an interval of space/time $(\overline{S_A}, \overline{\triangle t_A})$ with which both such systems are congruent:*

$$\forall_{N,M}, \exists \overline{I_{\sigma_N \Leftrightarrow \sigma_M}} \implies \exists (\overline{S_A}, \overline{\triangle t_A}) \mid (\sigma_N \in \overline{\sigma_{S_A, \triangle t_A}}) \& (\sigma_M \in \overline{\sigma_{S_A, \triangle t_A}}) \tag{34}$$

Any interaction has both a structural and a functional component. *For all N, M the interaction $\overline{I_{\sigma_N \Leftrightarrow \sigma_M}}$ between systems σ_N and σ_M is equal to a structural component $I_{St(\sigma_N \Leftrightarrow \sigma_M)}$ and a functional component $I_{F(\sigma_N \Leftrightarrow \sigma_M)}$:*

$$\forall_{N,M}, \overline{I_{\sigma_N \Leftrightarrow \sigma_M}} \equiv I_{St(\sigma_N \Leftrightarrow \sigma_M)} + I_{F(\sigma_N \Leftrightarrow \sigma_M)} \tag{35}$$

Therefore specifically for the two systems σ_1 and σ_2 of interest here, there is some region of space and time that they are both congruent with (which we labeled above as $(\overline{S_j}, \overline{\triangle t_j})$, and this interaction is equal to a structural and a functional component.

$$\exists \overline{I_{\sigma_1 \Leftrightarrow \sigma_2}} \Longrightarrow (\sigma_1 \cong (\overline{S_j}, \overline{\triangle t_j}))\&(\sigma_2 \cong (\overline{S_j}, \overline{\triangle t_j}))$$
$$\&(\overline{I_{\sigma_1 \Leftrightarrow \sigma_2}} = I_{St(\sigma_1 \Leftrightarrow \sigma_2)} + I_{F(\sigma_1 \Leftrightarrow \sigma_2)}) \quad (36)$$

(b) **The functional interaction between the two systems is repetitive and self-perpetuating, because the two systems are sufficiently similar in terms of how their respective functional rules relate to each other.** The interaction operates over a period of time consistent with the functional repetition of each system. In the case of functional similarity between the two systems, there is a continuous relation between the function of one system and that of the other. *There exists a relation $F_{SIMILAR(1,2)}$ between the function of systems σ_1 and σ_2, such that the function of each system is continuously and consistently related to that of the other through this relation, such that the interaction is sufficiently symmetrical that it continues to exist over time. This similarity relation is a subset of the interaction considered as a whole:*

$$\exists F_{SIMILAR(1,2)} \subset I_{F(\sigma_1 \Leftrightarrow \sigma_2)}) \mid \overline{F_{\sigma_1}} \quad \underset{F_{SIMILAR(1,2)}}{\overset{F_{SIMILAR(1,2)}}{\overset{\longrightarrow}{\longleftarrow}}} \quad \overline{F_{\sigma_2}} \quad (37)$$

This means that the systems are similar enough to each other that a consistent, ordered relationship can exist between them.

i. **Functional similarity between two systems means that the connection pattern linking particular causal rules within the function of one system, has properties in common with the connection pattern linking causal rules in the function of the other system.** The functional relation between the two systems $F_{SIMILAR(1,2)}$ preserves those properties that both systems have in common. The functional repetition of system one becomes "tied in" to that of system two, because there is a degree of similarity about how the repetition works in each system. This similarity is only a subset of the function of each individual system. The similarity between the two systems is a connection pattern, which amounts to what the self-referential functional repetition of each system shares with the other. This similarity is symmetric and commutative (independent of the order), so system one can influence system two through relation $F_{SIMILAR(1,2)}$, or system two can influence system one through the same relation). *The operation of the function of system one, with all its self-referential properties* $\overline{F_{\sigma_1}}^{\circlearrowleft\overline{\alpha_{\sigma_1}}}$, *followed by the function-function relation between the two systems* $F_{SIMILAR(1,2)}$, *followed by the operation of the function of system two, with all its self-referential properties* $\overline{F_{\sigma_2}}^{\circlearrowleft\overline{\alpha_{\sigma_2}}}$, *gives the same result as the interaction working the opposite way around, beginning with system two and ending with system one. This end result is equal to a connection pattern* $\overline{H_{\sigma_1,\sigma_2}}$, *which is a property of the connections between causal rules within the function of both system one and system two. This connection pattern can be thought of as roughly equivalent to the intersection* $\overline{F_{\sigma_1}} \cap \overline{F_{\sigma_2}}$ *between the functions of each system:*

$$\overline{F_{\sigma_2}}^{\circlearrowleft\overline{\alpha_{\sigma_2}}} \circ F_{SIMILAR(1,2)} \circ \overline{F_{\sigma_1}}^{\circlearrowleft\overline{\alpha_{\sigma_1}}}$$

$$= \overline{F_{\sigma_1}}^{\circlearrowleft\overline{\alpha_{\sigma_1}}} \circ F_{SIMILAR(1,2)} \circ \overline{F_{\sigma_2}}^{\circlearrowleft\overline{\alpha_{\sigma_2}}}$$

$$= \overline{H_{\sigma_1,\sigma_2}} \approx \overline{F_{\sigma_1}} \cap \overline{F_{\sigma_2}} \tag{38}$$

ii. **The fact that the connection patterns within the function of each system have this similarity between them, causes the interaction to be repetitive and stable over time.** The similarity between the connection patterns within the function of each system allows them to cycle through a kind of second-order repetition, such that the interaction between the two systems does not break down. *The existence of functional similarity* $\overline{H_{\sigma_1,\sigma_2}}$ *implies that for all time points within the time-frame* $\overline{\triangle t_A}$ *within which the interaction happens, the operation of the function* $\overline{F_{\sigma_1}}$ *of system one, combined with its self-referential functional state* $\overline{\alpha_{\sigma_1}}$, $\overline{F_{\sigma_1}}^{\circlearrowleft\overline{\alpha_{\sigma_1}}}$, *and the equivalent operation of system two and its functional state* $\overline{F_{\sigma_2}}^{\circlearrowleft\overline{\alpha_{\sigma_2}}}$, *in the presence of the interactive connection* $F_{SIMILAR(1,2)}$, *is equivalent to that same interaction at the previous time-point.*

$$\exists\overline{H_{\sigma_1,\sigma_2}} \Longrightarrow \forall t \in \overline{\triangle t_A}.\, \Big(\overline{F_{\sigma_1}}^{\circlearrowleft\overline{\alpha_{\sigma_1}}} \underset{F_{SIMILAR(1,2)}}{\overset{F_{SIMILAR(1,2)}}{\rightleftarrows}} \overline{F_{\sigma_2}}^{\circlearrowleft\overline{\alpha_{\sigma_2}}}\Big)_{(t)}$$

$$\equiv \Big(\overline{F_{\sigma_1}}^{\circlearrowleft\overline{\alpha_{\sigma_1}}} \underset{F_{SIMILAR(1,2)}}{\overset{F_{SIMILAR(1,2)}}{\rightleftarrows}} \overline{F_{\sigma_2}}^{\circlearrowleft\overline{\alpha_{\sigma_2}}}\Big)_{(t+1)} \tag{39}$$

Importantly, because this functional aspect of the interaction between the two systems is stable and self-perpetuating, each system continues to get exposed to the structure of the other, despite it being uncorrelated with that structure.

(c) **The two systems are fundamentally structurally different, with no consistent relation existing between the function of one system and the structure of the other, causing the structure of each system to influence the function of the other in an inconsistent, chaotic way.** In terms of the functional repetition of each system, the structure of the other system is "off the scale"; the rules operating in the structure of one of the interacting systems at any given point in time have no systematic relationship to the rules operating in the function of the other system at that same time-point. The impact that the structure of one system has on the function of the other is therefore chaotic, meaning that its nature and net consequences are not consistent over time. *Consider the effect of the structure of system one* $\overline{St_{\sigma_1}}$ *on the function of system two* $\overline{F_{\sigma 2}}$ *(with its repetitive self-referential properties* $\overline{\alpha_{\sigma_2}}$*). Label this effect* $r_{\overline{(St_{\sigma_1} \Longrightarrow F_{\sigma_2})}}$ *(the relation between the structure of system one and the function of system two), such that the structure of system one* $\overline{St_{\sigma_1}}$ *is related to the function of system two* $\overline{F_{\sigma_2}}$ *via this relation:*

$$\exists r_{\overline{(St_{\sigma_1} \Longrightarrow F_{\sigma_2})}} \mid \overline{St_{\sigma_1}} \overset{r_{\overline{(St_{\sigma_1} \Longrightarrow F_{\sigma_2})}}}{\longrightarrow} \overline{F_{\sigma_2}}^{\circlearrowleft \overline{\alpha_{\sigma_2}}} \tag{40}$$

The overall result of the simultaneous presence of the structure of system one, the function of system two, and the relation between them, is not the same at one time point as it was at any previous time point. This is the key "chaotic" property of the structure to function influence. *For all* t *within the interaction time interval* $\overline{\Delta t_A}$*, the operation of the structural rules of system one* $\overline{St_{\sigma_1}}$*, followed by the structure to function relation* $r_{\overline{(St_{\sigma_1} \Longrightarrow F_{\sigma_2})}}$*, followed by the functional rules of system two* $\overline{F_{\sigma_2}}$ *(i.e. the sequence* $\overline{St_{\sigma_1}} \circ r_{\overline{(St_{\sigma_1} \Longrightarrow F_{\sigma_2})}} \circ \overline{F_{\sigma 2}}$*) does not produce the same overall consequences at time point* t *as it does at the subsequent time point* $t+1$*. This implies that all causal rules contained within the structure to function relation are chaotic (non-repetitive and inconsistent in time)* $\forall j \mid r_j \in r_{\overline{(St_{\sigma_1} \Longrightarrow F_{\sigma_2})}}, r_j \in \overline{R_K}$*:*

$$\forall t \in \overline{\Delta t_A}, (\overline{St_{\sigma_1}} \circ r_{\overline{(St_{\sigma_1} \Longrightarrow F_{\sigma_2})}} \circ \overline{F_{\sigma 2}})_{(t)}$$

$$\neq (\overline{St_{\sigma_1}} \circ r_{\overline{(St_{\sigma_1} \Longrightarrow F_{\sigma_2})}} \circ \overline{F_{\sigma 2}})_{(t+1)} \Longrightarrow \forall j \mid r_j$$

$$\in r_{\overline{(St_{\sigma_1} \Longrightarrow F_{\sigma_2})}}, r_j \in \overline{R_K} \tag{41}$$

The interaction between the two systems is symmetrical; the exact equivalent argument holds for the interaction between the structure of system two $\overline{St_{\sigma_2}}$, the function of system one $\overline{F_{\sigma_1}}$ and the (chaotic) relation connecting them $r_{\overline{(St_{\sigma_2} \Longrightarrow F_{\sigma_1})}}$:

$$\exists r_{\overline{(St_{\sigma_2} \Longrightarrow F_{\sigma_1})}} \mid \overline{St_{\sigma_2}} \overset{r_{\overline{(St_{\sigma_2} \Longrightarrow F_{\sigma_1})}}}{\longrightarrow} \overline{F_{\sigma_1}}^{\circlearrowleft \overline{\alpha_{\sigma_1}}} \tag{42}$$

$$\begin{aligned} &\forall t \in \overline{\Delta t_A}, (\overline{St_{\sigma_2}} \circ r_{\overline{(St_{\sigma_2} \Longrightarrow F_{\sigma_1})}} \circ \overline{F_{\sigma_1}})_{(t)} \\ &\neq (\overline{St_{\sigma_2}} \circ r_{\overline{(St_{\sigma_2} \Longrightarrow F_{\sigma_1})}} \circ \overline{F_{\sigma_1}})_{(t+1)} \\ &\Longrightarrow \forall j \mid r_j \in r_{\overline{(St_{\sigma_2} \Longrightarrow F_{\sigma_2})}}, r_j \in \overline{R_K} \end{aligned} \tag{43}$$

Thus, the function of each system is continually exposed to a chaotic input as a result of the structure of the other system. Despite its chaotic nature, this input persists because the function to function interaction between the systems (described in the previous point) is repetitive and self-perpetuating - as a result of the similarity between the functions of each system. In the case of some interactions, the self-perpetuating functional interaction may not be sufficient to keep the interaction going, and some change induced by the chaotic influence of the other system may cause the interaction to break down, and/or one of the systems to be destroyed. However, sometimes the interaction will continue, and in such circumstances a prerequisite step for novelty will have been passed.

(d) **The function of each system gains changes as a consequence of the chaotic interaction with the structure of the other system. These "extra" changes arise from the interaction between the functional changes and the associated rules, and the chaotic effect of the other system's structure.** The continued presence of the structure of each system causes random changes to happen as it impacts upon the function of the other system. Some of these random changes are associated with each other, i.e. they are linked by causal rules. But these changes, and any causal rules describing their association, are not repetitive. Therefore nor is the system's functional response. *The relation $r_{\overline{(St_{\sigma_1} \Longrightarrow F_{\sigma_2})}}$ describing the impact of the structure of system one on the function of system two, is equivalent to a series of induced changes $\sum_j C_j$, each one of which is linked to another whole sequence $\sum_k C_k$ of subsequent changes (the length and nature of which depends on the details of the system and the nature of* *the prior change it is connected to). (I.e. sequence* $(\sum_k C_{k(t+1)})_j$ *of* k *subsequent changes, caused by and specific to the particular change* C_j*). Each connection between these extra, structurally induced changes is described by a causal rule* $r_{C_{j,k}}$*, which in all cases (for all* $\forall j, k$ *such that causal rule* $r_{C_{j,k}}$ *links structurally induced changes* C_j *and* C_k*,i.e.*$C_{j(t)} \overset{r_{C_{j,k}}}{\longrightarrow} C_{k(t+1)}$*), is inconsistent in its operation over time, i.e. a member of the category chaos* $\overline{R_K}$, $\forall j, k, r_{C_{j,k}} \in \overline{R_K}$ *. Any such rule contained within the structure to function relation category is not a member of the function of the system it is influencing. E.g. any rule elicited by the structure of system one on the function of system two, is not part of the function of system two* $\forall j, k, r_{C_{j,k}} \notin \overline{F_{\sigma_2}}$*. This implies that the operation of the structure to function relation from system one to system two* $\forall r_{C_{j,k}} \in r_{\overline{(St_{\sigma_1} \Longrightarrow F_{\sigma_2})}}$ *over time* $\overline{F_{\sigma 2(t)}} \overset{r_{C_{j,k}}}{\longrightarrow} \overline{F_{\sigma 2(t+1)}}$*, is also chaotic. The operation of the function of the system being impacted upon by any such rule* $r_{C_{j,k}}$ *within this structure to function relation, is therefore also partly chaotic: the* l *operation of the function after being impacted upon by this rule* $\overline{F_{\sigma 2(t+1)}}^{\circlearrowleft \overline{\alpha_{\sigma 2}}} \circ r_{C_{j,k}}$ *is therefore not equal to its previous operation* $\overline{F_{\sigma 2(t)}}^{\circlearrowleft \overline{\alpha_{\sigma 2}}}$*:*

$$r_{\overline{(St_{\sigma_1} \Longrightarrow F_{\sigma_2})}} \equiv \sum_j C_{j(t)}$$

$$| \sum_j C_{j(t)} \overset{r_{C_{j,k}}}{\longrightarrow} (\sum_k C_{k(t+1)})_j \mid \forall j, k, (r_{C_{j,k}} \in \overline{R_K}, \notin \overline{F_{\sigma_2}})$$

$$\Longrightarrow \forall r_{C_{j,k}} \in r_{\overline{(St_{\sigma_1} \Longrightarrow F_{\sigma_2})}}, \overline{F_{\sigma 2(t)}} \overset{r_{C_{j,k}}}{\longrightarrow}$$

$$\overline{F_{\sigma 2(t+1)}} \mid \overline{F_{\sigma 2(t+1)}}^{\circlearrowleft \overline{\alpha_{\sigma 2}}} \circ r_{C_{j,k}} \neq \overline{F_{\sigma 2(t)}}^{\circlearrowleft \overline{\alpha_{\sigma 2}}} \tag{44}$$

Again, the exact equivalent arguments are true for the relationship in the other direction; the impact of the structure of system two on the function of system one:

$$r_{\overline{(St_{\sigma_2} \Longrightarrow F_{\sigma_1})}} \equiv \sum_j C_{j(t)} \overset{r_{C_{j,k}}}{\longrightarrow} (\sum_k C_{k(t+1)})_j$$

$$| \forall j, k, (r_{C_{j,k}} \in \overline{R_K}, \notin \overline{F_{\sigma_1}})$$

$$\Longrightarrow \forall r_{C_{j.k}} \in r_{\overline{(St_{\sigma_2} \Longrightarrow F_{\sigma_1})}}, \overline{F_{\sigma_{1(t)}}}$$

$$\xrightarrow{r_{C_{j.k}}} \overline{F_{\sigma_{1(t+1)}}} \mid \overline{F_{\sigma_{1(t+1)}}}^{\circlearrowleft \overline{\alpha_{\sigma_1}}} \circ r_{C_{j.k}} \neq \overline{F_{\sigma_{1(t)}}}^{\circlearrowleft \overline{\alpha_{\sigma_1}}} \quad (45)$$

To put the argument in a crude form, extra changes happen due to the presence of the "alien" structure of the other system. These changes might, in other contexts be connected to the causation involved in the structure of this other system, but in the context of the functional interaction, their causal association with any other changes is chaotic; inconsistent in time.

(e) **The presence of the functional interaction between the two systems, by providing a partial link to the structure of the other system, causes some of the chaotic causal rules induced by this structural interaction to become transiently repetitive.** So, we have chaotic changes, induced upon the function of each system by the structure of the other. But simultaneously, the presence of the (ordered, repetitive) functional interaction, means that there is an indirect relationship that the function of one system has to the structure of the other. This causes there to exist intermittent periods of time during which the rules associated with changes induced by the "foreign" structure become transiently repetitive. Say the structure of system one induces chaotic changes in the functional repetition of system two. If system two were not functionally similar to system one, that would be the end of the matter. But the functional similarity is sufficient to partly "translate" the structure of system one to the structure of system two, for brief intervals during the interaction. This translation doesn't persist in time because it depends on how the structure of system one is actually affecting system two, which is not consistent in time. The functional similarity between the two systems causes the impact of the extra, structurally induced changes, to interchange between being chaotic and being repetitive. *With respect to system one σ_1, the existence of the functional similarity $F_{SIMILAR(1,2)}$ between the two systems implies that there exists a time point t_{NEW,σ_1} which is specific to the functional repetition of system one. The effect upon system one of the structure of system two is such that there exists a chaotic causal rule $r_{j,k} \in \overline{R_K}$, which operates upon the function of system one, as a result of the presence of the structure of system two $r_{j,k} \in r_{\overline{(St_{\sigma_2} \Longrightarrow F_{\sigma_1})}}$. This rule is such that it operates (the rule is an event, an instance of the operation of a specific causal rule at a specific point in time) at time point t_{TR,σ_1}, i.e. $(r_{j,k}(t_{TR.\sigma_1}) \in \overline{E})$. It is also true that the functional repetition of system one $\overline{F_{\sigma_1}}^{\circlearrowleft \overline{\alpha_{\sigma_1}}}$, and the functional repetition of system two*

$\overline{F_{\sigma_2}}^{\circlearrowright\overline{\alpha_{\sigma_2}}}$, *combined with the interactive functional similarity relation between them* $\overline{F_{\sigma_2}}^{\circlearrowright\overline{\alpha_{\sigma_2}}} \circ F_{SIMILAR(1,2)} \circ \overline{F_{\sigma_1}}^{\circlearrowright\overline{\alpha_{\sigma_1}}}$, *operates at this time point. This combined operation is impacted upon by the various causal rules within the chaotic relation between the structure of system two* $\overline{St_{\sigma_2}}$ *to the function of system one* $r_{\overline{(St_{\sigma_2} \Longrightarrow F_{\sigma_1})}}$, *and by the structure to function relation* $r_{\overline{(St \Longrightarrow F)}_{\sigma_2}}$ *within system two. Crucially, the combined, connected operation (which is written as the sequence* $(\overline{F_{\sigma_2}}^{\circlearrowright\overline{\alpha_{\sigma_2}}} \circ F_{SIMILAR(1,2)} \circ \overline{F_{\sigma_1}}^{\circlearrowright\overline{\alpha_{\sigma_1}}} \circ (r_{\overline{(St_{\sigma_2} \Longrightarrow F_{\sigma_1})}} + r_{\overline{(St \Longrightarrow F)}_{\sigma_2}}) \circ \overline{St_{\sigma_2}})_{(t_{TR,\sigma_1})})$ *of all of the causal rules associated with these distinct categories, at time point* t_{TR,σ_1}, *causes the anomalous continued operation, at the next time point, of the chaotic causal rule* $r_{j,k}$

$$\exists F_{SIMILAR(1,2)} \Longrightarrow \exists t_{NEW,\sigma_1} \& \exists r_{j,k}$$
$$\mid (r_{j,k} \in r_{\overline{(St_{\sigma_2} \Longrightarrow F_{\sigma_1})}}, \in \overline{R_K}) \mid \& (r_{j,k}(t_{TR,\sigma_1}) \in \overline{E}) \&$$
$$((\overline{F_{\sigma_2}}^{\circlearrowright\overline{\alpha_{\sigma_2}}} \circ F_{SIMILAR(1,2)} \circ \overline{F_{\sigma_1}}^{\circlearrowright\overline{\alpha_{\sigma_1}}}$$
$$\circ (r_{\overline{(St_{\sigma_2} \Longrightarrow F_{\sigma_1})}} + r_{\overline{(St \Longrightarrow F)}_{\sigma_2}}) \circ \overline{St_{\sigma_2}}))_{(t_{TR,\sigma_1})}$$
$$\Longrightarrow r_{j,k}(t_{TR,\sigma_1} + 1) \in \overline{E} \tag{46}$$

This kind of repetition of quasi-chaotic causal rule $r_{j,k}$ does not persist in the limit of infinite time. The functional similarity between the two systems is only enough to cause transient repetition of the chaotic causal rule, not to cause the origin of genuine repetitive causal rules ("genuine" meaning in this context "potentially infinitely repetitive in the right circumstances"). *It is not true that there exists this sort of repetition for all time points as time t tends to infinity:*

$$\neg \exists t_{TR,\sigma_1} \forall t \to \infty \tag{47}$$

The exact same arguments apply the other way around; the interaction between the two systems is symmetrical. The structure of system one influences the function of system two in a chaotic way, but the functional similarity between the two systems means that some chaotic causal rules, resulting from this structural influence become transiently repetitive from a system-specific time point t_{TR,σ_2}:

$$\exists F_{SIMILAR(1,2)} \Longrightarrow$$
$$\exists t_{TR,\sigma_2} \& \exists r_{j,k} \mid (r_{j,k} \in r_{\overline{(St_{\sigma_1} \Longrightarrow F_{\sigma_2})}}, \in \overline{R_K}) \& (r_{j,k}(t_{TR,\sigma_2}) \in \overline{E}) \&$$
$$((\overline{F_{\sigma_1}}^{\circlearrowright\overline{\alpha_{\sigma_1}}} \circ F_{SIMILAR(1,2)} \circ \overline{F_{\sigma_2}}^{\circlearrowright\overline{\alpha_{\sigma_2}}}$$
$$\circ (r_{\overline{(St_{\sigma_1} \Longrightarrow F_{\sigma_2})}} + r_{\overline{(St \Longrightarrow F)}_{\sigma_1}}) \circ \overline{St_{\sigma_1}}))_{(t_{NEW,\sigma_2})}$$

$$\implies r_{j,k}(t_{TR,\sigma_2}+1)\in\overline{E} \tag{48}$$

$$\neg\exists t_{TR,\sigma_2}\forall t\to\infty \tag{49}$$

Note that there is no reason to suppose that these transient repetition time points t_{TR,σ_1} and t_{TR,σ_2} within the different systems to be equal to each other, they may or may not be, its is a matter of luck. We can define the set of time points over which transient repetition occurs for each system, such that for each time point in this set, this time point marks the onset of transient repetition of an initially chaotic, structurally induced causal rule. *There exists the set $\overline{t_{TR,\sigma_j}}$ (where $j=1$ for system one and $j=2$ for system two), such that this set contains time point t_1, then t_2, up to t_n, and all time points in this set $\forall t\in\overline{t_{TR,\sigma_j}}$ are such that there exists transient repetition of the relevant (chaotic, quasi-repetitive) causal rule at this time point:*

$$\exists\overline{t_{TR,\sigma_j}}=[t_1,t_2...t_n]\mid\forall t\in\overline{t_{TR,\sigma_j}},(\exists(t_{TR,\sigma_j}))\implies(\exists(t_{TR,\sigma_j}+1)) \tag{50}$$

(f) **In some but not all of the interactions in which both systems exhibit transient repetition, this results in the origin of a new structure to function relation, which operates at the level of both individual systems but is not reducible to either independently (i.e. is emergent). The onset of such an emergent structure-function relation amounts to causing some of the initially chaotic causal rules to become permanently (i.e. non-transiently) repetitive. There exists a time point $t_{NOVELTY}$ after which system one σ_1 and system two σ_2 collectively exhibit repetitive causal rules that did not previously exist. The first instance of the operation of these new, subsequently repetitive causal rules, is an instance of objective novelty.**

i. **Define a set containing all the time points at which transient repetition occurs in both systems.** *There exists a set $\overline{t_{TR(\sigma_1,\sigma_2)}}$, (which is a subset of the interaction time interval $\overline{\Delta t_A}$) such that for all time points t within this set, it is true that there exists transient repetition of causal rules connecting initially chaotic, structurally induced changes within both system one and system two $(\exists t_{TR,\sigma_1}\&\exists t_{TR,\sigma_2})_{(t)}$:*

$$\exists\overline{t_{TR(\sigma_1,\sigma_2)}}\subset\overline{\Delta t_A}\mid\forall t\in\overline{t_{TR(\sigma_1,\sigma_2)}}\implies(\exists t_{TR,\sigma_1}\&\exists t_{TR,\sigma_2})_{(t)} \tag{51}$$

ii. **Within this set of time points $\overline{t_{TR(\sigma_1,\sigma_2)}}$ it is possible to define a point at which novelty occurs, and from which point onwards a structure-function relation that had never previously existed comes into being.** *There* *exists a time point* $t_{NOVELTY}$, *which is a member of the set of time points* $\overline{t_{TR(\sigma_1,\sigma_2)}}$, *during which both systems are exhibiting transient repetition. The existence of this time point implies that there exists, for all subsequent time points, a structure to function relation at the level of both initial systems, which did not exist prior to this time point:*

$$\exists t_{NOVELTY} \in \overline{t_{TR(\sigma_1,\sigma_2)}} \Longrightarrow$$

$$(\exists r_{(\overline{St \Longrightarrow F})_{\sigma_{NOVEL}}}(t), \forall t \geq t_{NOVELTY})$$

$$\&(\nexists r_{(\overline{St \Longrightarrow F})_{\sigma_{NOVEL}}}(t), \forall t < t_{NOVELTY}) \tag{52}$$

iii. **This structure function relation describes a systematic relationship between a novel structure and a novel function, both of which emerge as a consequence of the interaction between the two systems.** *The existence of the structure to function relation* $r_{(\overline{St \Longrightarrow F})_{\sigma_{NOVEL}}}$ *implies that there exists a corresponding structure* $\overline{St_{\sigma_{NOVEL}}}$, *function* $\overline{F_{\sigma_{NOVEL}}}$, *and self-referential connections between functional causal rules* $\overline{\alpha_{\sigma_{NOVEL}}}$, *connected by the operation of this structure-function relation* $\overline{St_{\sigma_{NOVEL}}} \overset{r_{(\overline{St \Longrightarrow F})_{\sigma_{NOVEL}}}}{\longrightarrow} \overline{F_{\sigma_{NOVEL}}}^{\circlearrowleft \overline{\alpha_{\sigma_{NOVEL}}}}$. *This is, in effect, equivalent to saying that there exists an entirely new system* σ_{NOVEL}:

$$\exists r_{(\overline{St \Longrightarrow F})_{\sigma_{NOVEL}}} \Longrightarrow \exists \overline{St_{\sigma_{NOVEL}}}, \overline{F_{\sigma_{NOVEL}}} \mid$$

$$\overline{St_{\sigma_{NOVEL}}} \overset{r_{(\overline{St \Longrightarrow F})_{\sigma_{NOVEL}}}}{\longrightarrow} \overline{F_{\sigma_{NOVEL}}}^{\circlearrowleft \overline{\alpha_{\sigma_{NOVEL}}}} \equiv \exists \sigma_{NOVEL} \tag{53}$$

iv. **The structure of this new system is simply the sum of the structures of the two initial systems with which the interaction began.** *The structure of the new system* $\overline{St_{\sigma_{NOVEL}}}$ *is the sum* $\overline{St_{\sigma_1}} + \overline{St_{\sigma_2}}$ *of the structure of the two starting systems:*

$$\overline{St_{\sigma_{NOVEL}}} = \overline{St_{\sigma_1}} + \overline{St_{\sigma_2}} \tag{54}$$

The spatial domain of the new system is the sum of the spatial domains of the two starting systems. *The spatial region* $\overline{S_{NOVEL}}$ *with which the new system is congruent is the sum of the region of operation of system one* $\overline{S_1}$*, and that of system two* $\overline{S_2}$*, minus the intersection between these two regions (to avoid counting the region of overlap twice):*

$$\overline{S_{NOVEL}} = \overline{S_1} + \overline{S_2} - \overline{S_1} \cap \overline{S_2} \tag{55}$$

Both the structure and function of the new system are congruent with this region of space:

$$\overline{St_{\sigma NOVEL}} \cong \overline{S_{NOVEL}} \tag{56}$$

$$\overline{F_{\sigma NOVEL}} \cong \overline{S_{NOVEL}} \tag{57}$$

v. **All causal rules in the novel system, including those that previously did not exist or were chaotic and/or transiently repetitive, are fully repetitive.** *For all rules* $r_{j,k}$ *within the function of the new system* $r_{j,k} \in \overline{F_{\sigma NOVEL}}$*, this rule being part of the new system implies that it is (potentially infinitely, as opposed to transiently) repetitive* $r_{j,k} \in \overline{R_R}$ *, in the sense defined above; i.e. there exist changes* C_j, C_k *such that the connection between these two changes is described by rule* $r_{j,k}$*:*

$$\forall r_{j,k}, r_{j,k} \in \overline{F_{\sigma NOVEL}} \Longrightarrow$$

$$r_{j,k} \in \overline{R_R} \Longrightarrow \exists C_j, C_k \mid (C_{j(t)} \overset{r_{j,k}}{\longrightarrow} C_{k(t+1)} \forall t \tag{58}$$

vi. **The function of the novel system contains either causal rules that were in both initial systems, or entirely new causal rules that had not existed at any previous time point prior to the onset of transient repetition.** The function of the novel system $\overline{F_{\sigma NOVEL}}$ is a collection of repetitive causal rules (which may be linked together in a self-referential network $\overline{\alpha_{\sigma NOVEL}}$ as defined above), which are either present in the functional intersection $\overline{H_{\sigma_1,\sigma_2}}$ between the two initial systems, or are entirely new $\overline{r_{NEW}}$ (but obviously no causal rules are in both categories).

$$\overline{F_{\sigma NOVEL}} = \overline{H_{\sigma_1,\sigma_2}} + \overline{r_{NEW}} \tag{59}$$

For all rules $r_{j,k}$ *connecting the* j*th and* k*th changes, this rule being a member of the function of the new system* $\overline{F_{\sigma NOVEL}}$ *implies that either it is a member of the functional intersection between the two original systems* $r_{j,k} \in \overline{H_{\sigma_1,\sigma_2}}$*, or entirely new* $r_{j,k} \in \overline{r_{NEW}}$*, but not both* $\neg(r_{j,k} \in \overline{r_{NEW}} \& r_{j,k} \in \overline{H_{\sigma_1,\sigma_2}})$*:*

$$\forall r_{j,k}, r_{j,k} \in \overline{F_{\sigma NOVEL}} \Longrightarrow$$

$$(r_{j,k} \in \overline{H_{\sigma_1,\sigma_2}}) \vee (r_{j,k} \in \overline{r_{NEW}})\&$$

$$(\neg(r_{j,k} \in \overline{r_{NEW}}\& r_{j,k} \in \overline{H_{\sigma_1,\sigma_2}})) \quad (60)$$

The functional intersection between the two starting systems contains non-novel causal rules that were present in both the initial systems. *For all rules in the functional intersection between the two parent systems* $\forall r_{j,k}, r_{j,k} \in \overline{H_{\sigma_1,\sigma_2}}$, *the rule's being in the category* $\overline{H_{\sigma_1,\sigma_2}}$ *implies its being in both the function of system one* $r_{j,k} \in \overline{F_{\sigma_1}}$ *and that of system two* $r_{j,k} \in \overline{F_{\sigma_2}}$*:*

$$\forall r_{j,k}, r_{j,k} \in \overline{H_{\sigma_1,\sigma_2}} \implies (r_{j,k} \in \overline{F_{\sigma_1}})\&(r_{j,k} \in \overline{F_{\sigma_2}}) \quad (61)$$

The causal rules in the new category $\overline{r_{NEW}}$ initiated at the start of the new system, were not present in either of the starting systems, and may have been transiently repetitive during the chaotic part of the interaction. *For all new causal rules* $r_{j,k}, r_{j,k} \in \overline{r_{NEW}}$, *their membership of the category* $\overline{r_{NEW}}$ *implies their non-membership of the original function of either system one or system two* $r_{j,k} \notin \overline{F_{\sigma_1}}\& r_{j,k} \notin \overline{F_{\sigma_2}}$, *and that prior to time point* $t_{NOVELTY}$, *such rules had never previously operated repetitively* $\forall t < t_{NOVELTY}, r_{j,k} \notin \overline{R_R}$. *This is effectively equivalent to saying that such repetitive rules did not exist at all prior to this time point* $\nexists r_{j,k}(\forall t < t_{NOVELTY})$*:*

$$\forall r_{j,k}, r_{j,k} \in \overline{r_{NEW}} \implies r_{j,k} \notin \overline{F_{\sigma_1}}\& r_{j,k} \notin \overline{F_{\sigma_2}}$$

$$\&\forall t < t_{NOVELTY}, r_{j,k} \notin \overline{R_R} \equiv \nexists r_{j,k}(\forall t < t_{NOVELTY}) \quad (62)$$

vii. **The first instance of operation of those causal rules, associated with the new system, which were not present in either initial system, are instances of novelty - discontinuity in the distribution of causal rules within time.** *The existence of* $t_{NOVELTY}$ *implies that for all causal rules in new category* $\overline{r_{NEW}}$, *the operation of these causal rules at this specific time point is an instance of novelty* $r_{j,k}(t_{NOVELTY}) \in \overline{N}$, *and at all subsequent time points the operation of this rule is an event (i.e. the rule continues to operate). The rule in question was chaotic before* $t_{NOVELTY}$ $(r_{j,k}(t < t_{NOVELTY}) \in \overline{R_K}$ *and became repetitive after* $r_{j,k}(t \geq t_{NOVELTY}) \in \overline{R_R}$. *This change defines the first instance of operation of the rule as a novelty* $r_{j,k}(t_{NOVELTY}) \in \overline{N}$*:*

$$\exists t_{NOVELTY} \implies \forall r_{j,k}, r_{j,k} \in \overline{r_{NEW}}, (r_{j,k}(t_{NOVELTY}) \in \overline{N})$$

$$\&(r_{j,k}(t > t_{NOVELTY}) \in \overline{E})$$

$$\implies (r_{j,k}(t < t_{NOVELTY}) \in \overline{R_K}) \& r_{j,k}(t \geq t_{NOVELTY}) \in \overline{R_R} \quad (63)$$

A few footnotes:

A. **Some causal rules present in the initial system(s) are lost at the origin of the new system.** *The existence of a causal rule present in either initial system* $\underline{\exists r_{j,k}} \in \overline{F_{\sigma_1}} \vee \overline{F_{\sigma_2}}$ *but absent in the new system* $r_{j,k} \notin \overline{F_{\sigma_{NOVEL}}}$ *implies that there were causal rules present before the breaking point to novelty* $\exists r_{j,k}(t < t_{NOVELTY})$ *that are absent after this breaking point* $\nexists r_{j,k}(t \geq t_{NOVELTY})$*:*

$$\exists r_{j,k} \in \overline{F_{\sigma_1}} \vee \overline{F_{\sigma_2}} \mid r_{j,k} \notin \overline{F_{\sigma_{NOVEL}}} \implies$$

$$(\exists r_{j,k}(t < t_{NOVELTY})) \& (\nexists r_{j,k}(t \geq t_{NOVELTY})) \quad (64)$$

In other words, novelty involves loss of causal rules as well as their origin. This has the implication that the only reason to expect complexity to accumulate as time passes is a statistical one, relying on the assumption that most instances of novelty will create more causal rules than they remove. This assumption, if correct, is only correct by accident, not by necessity. Although this does seem to be borne out in the history of life (for instance the proposed tendency for biological complexity to increase over time, "biology's first law"[14]), it is well worth noting that there is no fundamental necessity for complexity to increase, at least not as a result of novelty.

B. **There is no way to predict the interval of time with respect to which the new system is defined, on the basis of the interval(s) of time with respect to which the starting two systems were defined. This is because this new time interval involves functional repetition of causal rules that did not exist in either starting system.** *There does not exist any functional relationship* $f_{NOVEL,1,2}$ *such that the timescale congruent with the two initially interacting systems one* $\sigma_1 \cong \overline{\Delta t_1}$ *and two* $\sigma_2 \cong \overline{\Delta t_2}$ *can be related to the timescale congruent with the new system to which they gave rise* $\sigma_{NOVEL} \cong \overline{\Delta t_{NOVEL}}$.

$$\nexists f_{NOVEL,1,2} \mid (\sigma_1 \cong \overline{\Delta t_1}, \sigma_2 \cong \overline{\Delta t_2})$$

$$\&(\overline{\Delta t_1}, \overline{\Delta t_2} \overset{f_{NOVEL,1,2}}{\longrightarrow} \overline{\Delta t_{NOVEL}})$$

$$| \ \sigma_{NOVEL} \cong \overline{\Delta t_{NOVEL}} \tag{65}$$

In other words, knowing the timescale congruent with the two starting systems need not give you any information about the timescale congruent with the new system. This distinction between time and space arises because novelty is about the origin of causal rules that repeat in time, rather than about arrangements of stuff in space that can be frozen at a fixed point in time. The structure of the new system is the sum of the starting two, and thus obviously can be calculated with ease - whereas, by contrast, the time scale over which the new system persists can only be observed retrospectively.

C. **It is not true that there exists a breaking point leading to novelty in all interactions between functionally similar but structurally different systems, because not all such interactions will give rise to the above sequence of events.** *It is not true that the existence of a breaking point to novelty $\exists t_{NOVELTY}$ exists for all interactions between functionally similar and structurally different systems, for all j, k such that there exists an interaction between system j and system k.*

$$\neg \exists t_{NOVELTY(j,k)} \forall j, k \ | \ \exists \overline{I_{\sigma_j \Leftrightarrow \sigma_k}} \tag{66}$$

This includes those systems that exhibit simultaneous transient repetition. The existence of such transient repetition is necessary but not sufficient for novelty to exist. *It is not true that the existence of a set of time points during which both interacting systems are exhibiting transient repetition $\overline{t_{TR(\sigma_1, \sigma_2)}}$ necessarily implies that there exists a breaking point giving rise to novelty $\exists t_{NOVELTY}$ within this interaction:*

$$\neg (\exists \overline{t_{TR(\sigma_1, \sigma_2)}} \implies \exists t_{NOVELTY}) \tag{67}$$

All that can be done is identifying necessary conditions for novelty. The identification of conditions sufficient for novelty in all interactions is impossible, which goes hand in hand with the above statement that novelty is only identifiable retrospectively. There is no way to predict novelty outright, all that can be done is to explain it retrospectively after the fact, and compare situations in which it has occurred, by developing a "recipe of events" of the sort given here. There is no systematic set of rules by which the collection of systems that are in existence at the current time point can be related to those that will come into existence at the next time point.

D. **Within the between-system interactions in which novelty does not occur, the interaction between the two systems eventually breaks down in the limit of infinite time** (for the same kind of random reasons that caused it to begin). *The non-existence of a breaking point giving rise to novelty* $t_{NOVELTY(j,k)}$ *in an interaction* $\overline{I_{\sigma_j \Leftrightarrow \sigma_k}}$ *between systems* σ_j *and* σ_k *implies the non-existence of this interaction in the limit of infinite time*$\nexists \overline{I_{\sigma_j \Leftrightarrow \sigma_k}}(t \rightarrow \infty)$*:*

$$\nexists t_{NOVELTY(j,k)} \Longrightarrow \nexists \overline{I_{\sigma_j \Leftrightarrow \sigma_k}}(t \rightarrow \infty) \qquad (68)$$

Notes

1. Hawley, K. *Temporal Parts.* (Stanford Encyclopedia of Philosophy, 2010).
2. Lawvere, F.W. & Schanuel, S. *Conceptual Mathematics: A first introduction to Categories.* (Cambridge, 2009).
3. Smullyan, R. & Fitting, M. *Set Theory and the Continuum Problem.* (Dover, 2010).
4. Russell, B. *Introduction to Mathematical Philosophy.* (Routledge, 1993 (First published 1919)).
5. Lao Tzu. *The Tao Te Ching.* ~6th century B.C. http://plato.stanford.edu/entries/laozi/
6. Smullyan & Fitting. Ibid ref. 3.
7. Wittgenstein, L. *Tractatus Logico Philosophicus.* (Annalen der Naturphilosophie (W. Ostwald). 1921).
8. Bergson, H. *Time and Free Will: An essay on the immediate data of consciousness.* (Dover. 2001 (First published1910)).
9. Hume, D. *An enquiry concerning Human Understanding.* (Hackett 1993 (First published 1748)).
10. Jung, C. *Synchronicity: A acausal connecting principle.* (Princeton, 1960).
11. Sheldrake, R. *A new science of life: The hypothesis of morphic resonance.* (Park Street,1995).
12. Bohm, D. *Wholeness and the Implicate Order.* (Routledge, 1980).
13. Feyerabend, P. *Farewell to reason.* (Verso, 1987).
14. McShea, D. & Brandon, R. *Biology's First Law.* (Chicago, 2010).

THEORETICAL CONSIDERATIONS RELATING TO REAL INSTANCES OF NOVELTY DURING THE HISTORY OF LIFE ON EARTH

The goal of this chapter is to provide justification for the pragmatic assertion that novelty, discontinuity in the distribution of cause-effect relationships over time, occurs as a result of the interaction between systems that function in similar ways but are made from different stuff. Most of the issues that will be discussed relate to Earth, life and consciousness. These, I argue, are the three most novel natural phenomena, and they are on some level inseparable from each other. Life's inherent novelty is self evident from both the variety of form and function in the living world, and from the fact that life continuously changes the biogeochemical feedbacks that operate within the Earth system. Life has introduced qualitatively new systems of cause and effect into Earth's climate, and as a result the Earth is the most novel (known) planet. This biogeochemical novelty is ultimately attributable to the origin of heritable variation in fitness, which, in turn, has arisen via both direct changes in the hereditary material, and through symbiogenesis. Further (and more tentatively), it is possible to speculate that humans are the most novel form of life, precisely because human consciousness clearly has an unprecedented capacity for envisaging the hypothetical. Although consciousness is probably the most problematic natural phenomenon to define in a concrete manner, what can be said is that it does seem to be inherently tied to not just problem solving, but creative thought. Life, consciousness, and planet Earth are all wrapped up together so as to produce an inherently novel entity. Why? What properties distinguish life and consciousness from the other aspects of nature?

If there is a single central theme that will become apparent, it is that nature is intrinsically symbiotic, that novelty occurs through the interactive combination between qualitatively different systems (i.e. as opposed to the "unfolding" of patterns of causation that are reducible to a single efficient cause). This combination happens between entities with an inherently compatible temporal function but different structures that reflect different histories. Life, Earth, and consciousness each, in different ways, combine things that "should not" be together in terms of the rules by which they function. This chapter is an attempt to make this claim more concrete. Some additional general suggestions which are, I argue, of some relevance to the history of life from a natural science perspective, are as follows:

1. *The history of life on Earth clearly supports the basic claim that "novelty happens when systems that function in similar ways but are composed of*

different stuff interact in time and space."

2. *Life is defined by a duality between homeostasis and information transfer. Life's novelty is connected to the fact that homeostatic and information-transfer systems converge functionally, but are structurally different types of system and therefore introduce a randomness to each other.*
3. *To a large extent, the evolutionary function of DNA is not to generate novelty, but to assimilate the novelty that is generated physiologically by the interaction between organisms and environments in time and space. The origin of variation in fitness cannot be attributed to a discrete causal "atom of heredity" that arises as a consequence of random mutations in DNA sequence. Instead it is more probable that it originates by means of interactive combination between qualitatively different systems. Precisely how this happens remains uncertain, but it must involve symbiotic interactions and genetic assimilation of the physiological properties that arise either from these interactions, or from the response of organism physiology to extreme physical environments.*
4. *Significant evolutionary change is rare and saltatory, not ubiquitous and gradual, and transitions in biological complexity have resulted from the occurrence of extreme environments that make survival a more significant component of fitness than fecundity.*

In some cases the above claims are strongly supported by the available evidence. In others they involve a subjective "reading between the lines", due to the unavoidable coarseness and uncertainty involved with making inferences about deep time. The reader is asked to accept the speculative and hypothetical nature of this discussion in the open-minded, exploratory spirit in which it is intended. The arguments surrounding these issues are only described in sufficient detail to serve as a reference point for this central theoretical argument; the history of life on Earth is, obviously, an extremely detail-rich subject that would fill many books in its own right. (For a comprehensive discussion of the details of the kinds of new systems that have come into existence during the history of life, the reader may wish to consult one of the many excellent monographs relating to Earth system science and/or evolutionary biology. For instance the reader interested in Earth's climate processes might try Lovelock[1,2], Knoll[3], Volk[4], Schwartzman[5], Canfield[6], Lenton and Watson[7], or Kasting[8]. For evolutionary biology perhaps Crick[9], Maynard-Smith and Szathmary[10], Morowitz[11,12], Waddington[13], De Duve[14,15], or Margulis and Sagan[16,17]).

Life: nature's most novel process

The fact that life is more variable than the other processes of nature has been almost impossible to dispute since fossils and geological sciences first demonstrated that life has a history. This variability is connected to the inherent novelty generating capacity of life as a process, which in turn must surely be deep enough to relate to life's very definition. One such definition is "a system exhibiting simultaneous storage and replication of nucleic acid and enzyme catalysis"[18], although various others are conceivable (relating to various additional tangible properties of living things such as respiration, response to stimuli, feeding, movement, reproduction, etc). But defining life as the combination between enzyme catalysis and storage/replication of nucleic acid has the weight of practical utility behind it: A working scientist with appropriate modern equipment could reasonably conclude, without any room for ambiguity, that any system exhibiting both these properties was unequivocally alive, and no practical person would dispute his/her conclusion. Importantly, the definitive requirement for the simultaneous presence of two distinct features reflects how life literally is the duality between information and homeostasis.

Storage and replication of deoxyribose-nucleic acid, DNA[19] is the basic prerequisite for the information-based features of life. DNA is used by living cells to generate RNA (ribose-nucleic acid, a chemical variant of DNA), and this RNA is used to generate protein. At each specific instance at which this happens, the sequence of the DNA is copied, i.e. the sequence is transferred in the direction $DNA \longrightarrow RNA \longrightarrow Protein$. A nucleic acid is a polymer, a long molecule composed of many repeating phosphonucleotide subunits. Each nucleotide monomer consists of a nitrogen-derived nucleobase (adenine, thymine, cytosine or guanine), a central sugar molecule (ribose in the case of DNA, deoxyribose in the case of RNA) and a phosphate "backbone" (by means of which multiple nucleotides are polymerized together in a long chain that makes up a DNA strand). Each of the four bases can pair with one of the others (adenine with thymine, cytosine with guanine) by hydrogen bonds (charge based interactions that are more transient than other forms of chemical bond). As a result of this pairing, two DNA strands can bind together - the famous "double helix" ladder like structure discovered by Watson and Crick[20]. When cells make proteins, the DNA unwinds, and an enzyme complex known as RNA polymerase produces a strand of RNA, with a base pair sequence corresponding to the unwound DNA strand; a process labeled "transcription." This "messenger RNA" strand is then used by the cell to produce a protein, the sequence of which corresponds to the DNA sequence from which the messenger RNA is derived, i.e. $DNA \xrightarrow{transcription} RNA \xrightarrow{translation} Protein$[19]. A protein is a polymer of amino acids, biological molecules composed of nitrogenous amine $-NH_2$, carboxylic acid $-COOH$ functional groups, built around a side chain specific to each individual amino acid. A code exists whereby each amino acid that a cell uses has a triplet of RNA base pairs that corresponds to it, a sort of biochemical labeling system. A "transfer RNA" molecule binds, simultaneously, to both a specific amino acid, and section of 3 base pairs on the messenger RNA template molecule. While this binding holds the amino acids corresponding to each section of the messenger RNA

sequence in close proximity, other enzymes cause these amino acids to bind together permanently, with the end result being a polymer of amino acids, a protein, with sequence corresponding to the *DNA* sequence with which the whole process began, minus various non-coding and regulatory regions of the *DNA*, which do not make it in to the final, active, protein sequence, but may have some role in its developmental. Because proteins are responsible for much of the activity occurring within living cells, and because protein sequence has a significant effect on protein function, it is widely assumed that *DNA*, to a significant extent, controls cell function, therefore organism function, therefore organism reproduction. These facts, combined with the fact that the *DNA* sequence is, without fail, passed from parent to offspring, mean that there exists a reliable parent-offspring correlation in the disposition exhibited by organisms to survive and reproduce.

In order for it to operate within living populations, natural selection requires the existence of heritable variation and differential survival as a result of this heritable variation. More simply put: heredity, variation, selection. When all these pre-conditions are met, the variants within a population that are best adapted to surviving and reproducing in the particular natural environment will survive and reproduce better; i.e. have higher "fitness." Because variation is heritable, these "fitter" variants will make up a bigger fraction of the population in the next generation. Within an appropriate limit of time the fitter variants may "reach fixation" (take over the population). This, of course, is the basic logic of evolution by natural selection. These template-replication-driven, information-based properties of *DNA* sequences render variation in fitness heritable, and, as a result of this heredity, evolution by natural selection occurs. In other words, it is the information based properties of life that act as a foundation for biological evolution.

The strict dictionary definition of information is "The reception or communication of knowledge or intelligence"[21]. In this sense information can be thought of as the complementarity between an input and an interpretation of that input. In physics, there is a conceptual tie between information and entropy (granularity or disorder), which ultimately derives from the tendency for telecommunication signals to be lost during physical transmission processes[22]. In this (slightly different) physical/computational information concept, the "information-entropy" is the expected amount of information per message received, and a more improbable message-event provides more information. Biological entities do maintain an ordered (low entropy) internal environment, so in this sense there exists a loose parallel between biological information and thermodynamic entropy as granularity. But for the purposes of this discussion it is argued that biological information is best conceived of in terms of communication rather than entropy.

A book written in a language that one cannot understand contains no information. If one learns the language, the book suddenly contains information, even though the physical manifestation of the book (which some might relate to the book's internal "negative entropy" or degree of order) remains the same. Information derives not from the physical manifestation of the information-containing entity (in this metaphor, the paper the book is written on), but from the interaction between that entity and something capable of interpreting the information (the conscious mind reading the book). In exactly the same way, the information within biological systems is not reducible to the physical properties of a *DNA* molecule. Rather, biological information is a property of the interaction between a *DNA* molecule and the cellular machinery that gives rise to a working set of proteins. Nucleic acid contains information because biological systems are capable of "reading" such sequences and producing protein sequences from them, and because which parts of the *DNA* sequence are "read" depends on the specifics of protein activity (i.e. which proteins are required) in a living cell. Biological information is not "in" *DNA*. Rather, biological information is a relative property, it is the interactive, interpretative capacity of living cells to use *DNA* to make protein. The clearest and most biologically relevant definition of information is therefore a circular one; information is that which informs. Biological information is non-random sequence correspondence between qualitatively different macromolecules. For instance, there is no direct chemical similarity between amino acids and *DNA*; nucleotides and amino acids are chemically very different things; *DNA* and protein polymers are composed of different stuff. But they function in similar ways. The sequences of the two correspond to each other through what is literally a language-like syntax. Both *DNA* and protein are polymers of subunits that can be bound together so as to produce a sequence that is, in theory, arbitrary. Thus, one way of looking at the information-homeostasis duality, by which life has become the most variable part of the natural world, is as a symbiosis between nucleic acid and protein. Life derives its variety from the coupling together of macromolecules that are structurally very different, but have compatible consequences in terms of the chains of cause and effect with which they are associated.

Possession of nucleic acid is alone insufficient to render an entity genuinely alive. Viruses, most obviously, possess a nucleic acid genome, but they cannot replicate that genome unless they attack another living cell, hence cannot be considered alive in isolation from the host species upon which they depend. Other definitively living features are identifiable over shorter timescales than the information/replication based features of *DNA* template replication. These include response to stimuli, organization, metabolism, maintenance of an internal low entropy state, growth, movement, and so forth. Such features can be grouped under the umbrella term of homeostasis. The word homeostasis derives from two Greek words meaning "similar" (homeostatic systems target their composition towards a non-random "optimum" state that they are "aiming at") and "standing still" (homeostatic systems resist change imposed upon them from the outside). Living systems achieve homeostasis by the expenditure of energy. Consumption and transfer of energy, so as to maintain a low internal entropy state within a bounded micro-environment out of chemical equilibrium with the external environment, underlies most if not all alternative definitive[23] features of life. Such homeostatic control derives from the specificity of enzyme catalysis. An enzyme is a large protein molecule (or aggregate of protein molecules) within a living cell, which catalyzes (significantly increases the rate of) a specific chemical reaction (or set of reactions). Enzyme catalysis thus controls the occurrence of specific chemical reactions in time and space within living systems. This specificity reflects an inherent complementarity within biochemistry, which distinguishes it from the chemistry of non-living systems. Catalysis results from the fact that the enzyme molecule has a complementary shape to the reactant molecule. Enzymes bind with the "substrate" molecules that are required to react with each other, placing specific bonds under stresses that make the reaction more probable. The position on the (relatively large) enzyme molecule at which the (relatively small) substrate molecule binds is called the enzyme's "active site." The active site's shape precisely matches that of the substrate molecule - the famous "lock and key hypothesis"[19]. As Christian de Duve put it, biochemistry exhibits a fundamental complementarity that inorganic chemistry does not:

> *"The key notion here is complementarity between molecular configurations that fit and interlock with each other. This phenomenon, whose importance could hardly be overestimated, governs the combination of pieces in self-assemblies. It also explains the selection of substrates by the binding sites of enzymes. It is likewise involved in the interactions between active agents, such as hormones or drugs, with their cellular receptors, and in the recognition of antigens by antibodies in immune defense phenomena. We shall meet it as forming the basis of all genetic information transfers. Complementarity is often illustrated by the relationship between lock and key, or between mortise and tenon. The image is suggestive but only partly appropriate. Biological mortises and tenons have over those of cabinet makers the advantage of being more flexible and adaptable, so that they can to some extent mold themselves on each other. In addition, they bear with them, in the form of mutual affinities, the "glue" that helps stick them together."*[24]

Life's information based properties can thus be qualitatively distinguished from life's homeostasis based properties by their dynamic, self-referential qualities.

Importantly, this specific control over which chemical reactions happen when and where depends on maintaining chemical conditions that allow enzyme catalysis. This amounts to maintaining conditions that will allow particular proteins to keep their specific shape. Cell function goes awry when living cells are exposed to inhospitable temperature, pressure, pH and water-potential conditions, etc, because such conditions cause proteins to "denature" (lose their specific structure)[19]. Denaturing causes enzyme catalysis to fail, in turn causing the constituents of a living cell start to react in an uncontrolled manner, resulting in the cell to beginning to approach chemical equilibrium, i.e. to die. Avoiding such a situation requires spatial compartmentalization, and, particularly, the control of chemical reactions in time and space - i.e. enzyme catalysis. Thus, importantly, the conditions necessary to sustain enzyme catalysis are achieved through the effects of enzyme catalysis. This self-referential circularity is a definitive feature of any system exhibiting homeostatic control. That is, homeostasis is inherently teleological; i.e. purposeful. However, obviously, homeostatic "purposefulness" refers merely to the maintenance of a specific optimum state, and does not, of course, imply consciousness or sentience.

These properties reflect a teleological causation that Immanuel Kant[25] argued distinguishes living systems from non-living ones. Kant suggested that biology needs its own category of cause and effect, one that must be built around the notion that living things inherently possess a "natural purpose" that directs them towards a goal. In other words, for Kant, living things are inherently purposive, and this necessitates a legitimate teleological category of causation that distinguishes life from non-life. Importantly, this does not require some theological commitment as to the source of the teleology. Kant distinguished a hierarchical unidirectional cascade of discrete causes, from a bidirectional, holistic loop of causation, within which both cause and effect are connected:

> "*A causal connection, as our mere understanding thinks it, is one that always constitutes a descending series (of causes and effects): the things that are the effects, and that hence presuppose others as their causes, cannot themselves in turn be causes of these others. This kind of causal connection is called that of efficient causes (nexus effectivus). But we can also conceive of a causal connection in terms of a concept of reason (the concept of purposes). Such a connection considered as a series, would carry with it dependence both downwards and upwards: here we could call a thing the effect of something and still be entitled to call it, as the series ascends, the cause of that something as well.*"[25]

In other words, living things are definitively distinguished by the fact that they are both the cause and the effect of themselves. Kant's teleological definition of life relates to both efficient (immediate) and final (goal-directed) causes. It is therefore not enough to say that the constituent parts of a living thing share the same final cause (the goal of the organism). This could also be said of the constituent parts of a machine; which might share the same final cause with respect to the harmony necessary between their function required for that machine to work. This sort of machine-like interdependence between component parts, with respect to a final "output" also applies to the constituent parts of a living organism. But the parts of a living organism make themselves continuously over time, in contrast to the components of a designed machine. The constituent parts of living things are not only the final causes of each other with respect to their mutual interdependence for the purpose of the whole, but also, in some sense the immediate (efficient) causes of each other, through a process of self organization and growth. As McLaughlin[26] puts it in his analysis of Kant's thinking:

> *"....every part must be able to be viewed not only as the purpose or final cause of the others but also as the efficient or real cause of the production of the others, 'as an organ that produces the other parts', so that a natural purpose is an 'organized and self-organizing being'. Kant compares the natural purpose with a clock, whereby the major difference between the two lies in the fact that the parts of the clock may be conceived as final causes of one another with reference to the understanding that made the clock but not as efficient causes of one another. They do not give rise to one another and need not be so judged. An organized creature is more than a mere machine, because it has the power to form its parts and to transfer this formative power to the 'materials', so that the parts can mutually bring one another forth. Kant then points out that the analogy between artifact and organism is not worth much and even asserts that 'strictly speaking, therefore, the organization of nature has nothing analogous to any causality known to us'."*

Kant reached the conclusion that mechanistic (efficient causation based) thinking will never entirely explain biological systems, in large part connected to the elusive nature of their natural purposefulness. He therefore on the one hand argued that living things required a category of causation of their own, but on the other not only failed to define such a category himself, but stated that any such definition was impossible. This attitude culminated in Kant's famous statement that there will never arise a "Newton of the grass blade." Kant famously states that:

> *"For it is quite certain that in terms of merely mechanical principles of nature we cannot even adequately become familiar with, much less explain, organized beings and how they are internally possible. So certain is this that we may boldly state that it is absurd for human beings even to attempt it, or to hope that perhaps some day another Newton might arise who would explain to us, in terms of natural laws unordered by any intention, how even a mere blade of grass is produced."*

Francis Crick would be the clearest candidate for such a Newton of the grass blade, but teleological causation in biology is a property of homeostasis, not necessarily reducible to DNA in isolation. Self-referential, purposeful, teleological causation is entirely natural and amenable to scientific investigation, as opposed to being supernatural or requiring consciousness. Its definitive presence in biological systems is illustrated by the need to include enzyme catalysis in any useful definition of life.

Information and homeostasis

The inherent duality between information and homeostasis explains life's novelty generating capacity and underlies biological "chicken-egg" problems

A duality between information and metabolism is reflected in theoretical models seeking to explain the origin of life. Origin of life models are conceptually divided into "metabolism first" and "replicator first" (i.e. information first) scenarios[27]. Life must be able to reproduce itself, or else no evolution is possible and one cannot really call it life. But reproduction requires the persistence of an environment that permits the apparatus of reproduction to come into existence and be sustained, which requires metabolism and homeostasis. Creating and regulating such a non-random chemical environment is hard to imagine without enzyme catalysis, which in turn is hard to imagine without DNA. But homeostatic control of the cellular environment is necessary to sustain and reproduce the DNA. And so forth - self referentiality is built into the problem itself. This is what makes the origin of life the definitive chicken-egg question.

Information-first origin of life models emphasize replication. Once replicators arise, Darwinian evolution is possible, and once evolution can gain traction, the other features of life will follow more naturally. It is generally accepted that the most plausible "information first" origin of life scenarios centre around the notion of an "RNA-world"[28,29,30,31]. RNA is a nucleic acid sequence bearing information; RNA is chemically close to DNA, but RNA also has the capability of folding in upon itself so as to adopt a specific shape, like a protein, which allows RNA to exhibit some catalytic properties[29]. The RNA world scenario for the origin of life rests on the fact that RNA bridges the duality between catalytic control of biochemical reactions, and information transfer through sequence templates. The ribosome, a globular body of RNA involved in protein synthesis, has been suggested to be the direct descendent of a "ribozyme"[32] an autocatalytic (self-replicating) RNA molecule. However, net experimental replication of a specific RNA sequence, mimicking the origin of the first information replicators, requires the inclusion of an enzyme that cannot be produced by such RNA molecules alone. The catalytic properties of RNA alone are not enough to give rise to replication (i.e. an information-based origin of life) in the laboratory[33]. Furthermore, from a theoretical point of view, within the evolutionary dynamics of an imaginary a pool of self-replicating RNA molecules, a maximum sequence length would be reached, drastically constraining the potential for evolution within the RNA world, before such evolution really got started[33]. This is because at the replication of each base within an RNA sequence, there is a probability of error, therefore as the number of bases within a sequence increases, the probability of that sequence containing at least one error also increases. As the "genome size" of the self-replicating *RNA* molecule increases, the accumulation of errors becomes near catastrophic[34]. If it were assumed that this population of evolving *RNA* molecules was intermixed with an enzyme that improved replication accuracy, this problem would not arise, and *RNA* sequence length (genome size) could increase. But it is not possible to produce large, complex, enzyme molecules without the sequences of such molecules being encoded within a nucleic acid genome. As Maynard-Smith and Szathmary put it "*This is the catch 22 of prebiotic evolution: no enzymes without large genomes, no large genomes without enzymes.*"[35]. Although the duality of *RNA*'s information and catalytic properties does make the *RNA* world almost certainly part of the answer to the question of the origin of life, there is no getting away from the need for the combination of qualitatively different kinds of entity (enzymes and nucleic acid) to be forced together. The transition to novelty involved in the origin of life requires a forcing together of template replication and autocatalysis - different systems with nominal repetitive functional similarity.

Metabolism-first origin of life scenarios focus on a web of chemical reactions that is potentially self-perpetuating, known as an "autocatalytic loop." A chemical reaction or sequence of reactions is deemed autocatalytic if it regenerates the reactant molecules with which it began. Most realistically one must think in terms of a web of chemical reactions, that are collectively autocatalytic, and are contained in a physically discrete compartment that prevents the constituents diffusing away. Lipid membranes, like those of living cells, can form spontaneously due to the form adopted by fatty molecules when suspended in a water solvent[36]. The random enclosure of an autocatalytic web of reactions in a spherical lipid membrane "cell" is a reasonable basis for a metabolism first origin of life scenario. But in order to be stable, the reactions enclosed (randomly) in a compartment must be symmetrically autocatalytic: all the reactants in the loop must be regenerated in roughly equal proportions to those at the start[34] (or else the rate of various different steps in the chain will become imbalanced relative to the others). Even if something like a cell membrane could form spheres, and such spheres could split into smaller daughter spheres, any autocatalytic loop would not replicate stably on its own. An ordered cell division process is required, that replicates both the autocatalytic chemistry and the cell compartment. This, again, requires genetic material, raising another chicken-egg problem. Genomes do not spontaneously assemble and their replication requires chemical homeostasis. *DNA* polymerization requires energy, and *DNA* replication requires a cellular micro environment, with chemical conditions far from equilibrium with the exterior, within which *DNA* can be unwound, *RNA* assembled from it, proteins produced, the specificity and control of biochemistry achieved, and so forth. In other words, for a sufficiently complex network of reactions that would warrant the label biochemistry, one needs a relatively complex enzyme repertoire. For such a repertoire of enzymes one needs a DNA genome, transcription and translation from which is ordered in time and space. Of course, the constituent parts of enzymes, amino acids, can be generated abiotically, by the application of an electrical voltage to a pool consisting of simple abiotic chemicals such as methane, ammonia, water and hydrogen[37]: the famous Miller-Urey experiment. But this is a long way off generating the prerequisites for a complex metabolism, sufficient to catalyze and replicate an ordered, autocatalytic loop of chemical reactions in a physiologically controlled manner. There is no getting away from the need for including nucleic acid and protein together in models attempting to explain the origin of life[38].

This need for duality between qualitatively different kinds of system is thus generic and definitive within life as a natural phenomenon. Life consists of a combination of systems capable of information transfer and systems capable of homeostasis. Such systems necessarily function in compatible ways. Homeostasis achieves the persistence of the sorts of conditions that will allow replication/information-transfer. Information transfer allows longer term persistence of the apparatus by which that homeostasis is achieved. The chain of cause and effect is, exactly as Kant noted, both unique and inherently teleological and self-referential. But information transfer derives from the homology of sequences within $DNA \longrightarrow RNA \longrightarrow Protein$. This homology is not reducible to the chemistry of any of these molecules alone, it derives from the fact that there exists a non-random code[20], by which sequences are transferred between each of these molecules. Homeostatic control is not reducible to this code. Information and homeostasis "function in similar ways", in that chains of cause and effect relating to each property are compatible with each other. Information and homeostasis are "built from different stuff", in that the material constituents of nucleic acid replication and those of enzyme catalysis and physiological control are very different. Life's novelty generating capacity results from the fact that life is the interaction between information and homeostasis. This qualitative generalization is, of course, directly in accord with the idea of novelty from functional similarity and structural difference. This duality is why the definition of life, however formulated, must always be a list rather that a single property. It may also be connected to the possibility that developing a plausible origin of life scenario does not "just" reduce to getting the right set of chemical background conditions; the right combination of qualitatively different systems is also necessary.

The origin of heritable variation in fitness

If life is the most novel feature of nature, then the generation of novelty must be a central, definitive aspect of life as a natural phenomenon. This novelty generation, it is widely assumed, is ultimately attributable to random changes in DNA sequences and their evolutionary consequences. That nucleic acid sequence dictates protein sequence, with information passing in the direction $DNA \longrightarrow RNA \longrightarrow Protein$, and not the other way, is, of course, is Watson and Crick's[39] "central dogma" of molecular biology, one which has proven itself to be correct. Protein sequence determines protein shape, protein shape determines protein activity, protein activity determines everything else. Protein activity, via enzyme catalysis, controls the occurrence of chemical reactions in time and space, which ultimately drives changes in the form, function, and survival of living organisms. Yet it is a curious fact that even with modern computational power, knowledge of the amino acid sequence of a protein does not provide sufficient information to predict the structure of that protein[40]. This outstanding "protein folding problem" exists despite a generation of claims that modern bio science is only a few research grants away from complete predictive power over essentially all the qualities of living organisms. The latter attitude is illustrated most clearly by Lewis Wolpert's wager with Rupert Sheldrake, with Wolpert making the claim that "***By 1 May 2029, given the genome of a fertilized egg of an animal or plant, we will be able to predict in at least one case all the details of the organism that develops from it, including any abnormalities.***"[41].

Proteins are polymers of amino acids, which fold in on themselves through electrostatic interactions so as to adopt a complex three dimensional structure. Protein folding is hierarchical and self-referential, involving globular structures known as "folding domains"[19]. The existence of such folding domains causes electrostatic interactions to occur at the level of the shape of the domain in question, effectively forming a new electrostatic entity that modifies the future direction of the folding process at each stage. The shape of a protein sequence emerges over time, as the linear amino acid polymer (peptide sequence) folds in upon itself, creating new folding domains, that fold in upon themselves again to create a higher order set of constraints on folding, and so forth. Mathematicians are only able to "predict" protein shape in a probabilistic manner, through a search process in which a similar protein with a known shape is used as a reference point[42]. All that DNA sequences can do is modify amino acid sequences (or in some cases, the activity of other DNA sequences, which in turn modify amino acid sequences). But if amino acid sequences do not provide enough information to predict protein shape, let alone protein function, then what biochemical basis is there for supposing that random mutations in DNA sequence create the entirety of the variation in fitness upon which evolution has acted? How precisely do random changes in DNA produce changes in the form, function and fitness of living species?

It is informative as regards to the above question to note that the modern, theoretical concept of a "gene" pre-dates the discovery of DNA by over a generation. The "modern synthesis", upon which 20th century thinking in evolutionary biology was based, involved the recognition by R.A. Fisher[43,44] of the theoretical significance of the particulate inheritance process that had previously been suggested by the experiments of Gregor Mendel[45]. Mendel had shown that various qualities of pea plants were inherited as seemingly discrete packets, so that offspring plants had characteristics of either one or other of their parents, as opposed to characteristics intermediate between both (the debate was focused on complex organisms with two parents, rather than bacteria or other organisms capable of reproducing asexually). The presence of these discrete factors, these "atoms" (indivisible pieces) of heredity (correlation between parents and offspring), gives rise to the concepts of "dominance" and "recessiveness" (the presence of one dominant atom of heredity cancels out that of another recessive atom)[42]. Fisher (a mathematician working in 1918, well before the discovery of *DNA* in 1953) built a theory around the assumption that all biological inheritance occurred in the discrete manner suggested by Mendel's peas. Shortly after Mendel's work, pioneering developmental biologist Weisman[46] had developed the notion of a "germ plasm", a material determinative agency passed on from parents to offspring, necessary and sufficient to dictate the features of organisms, and passed on only through "germline" (reproductive) cells, not "somatic" (body) cells. In effect, Fisher's visionary work quantified Weisman's germ plasm concept, related it to Mendel's ideas about particulate inheritance, and reconciled both with Darwin's concept of evolution by natural selection[47].

The alternative to the assumption of Mendelian particulate inheritance is known as "blending inheritance", which refers to the idea that each offspring has characteristics that are the intermediate average of each of its parents. Blending inheritance raises conceptual problems in terms of evolution by natural selection. Variation in the form and function (phenotype) of living organisms dictates variation in survival and reproduction (fitness). If evolutionary change is driven by natural selection upon variation in fitness, and if this evolutionary change is continuous, gradual and constantly occurring, then variation in fitness must be consistent across generations - or else there is nothing upon which selection can act. For example, suppose there exists some selection pressure on (say) eye colour in humans; suppose brown eyed people survive and reproduce better than blue eyed people for some reason. Suppose eye colour is inherited by a mechanism of blending inheritance. Say a couple, one of whom has brown eyes, one blue, have a child. Under a system of blending inheritance, the child's eyes will be neither brown nor blue, but in between. If this happens across the population, then at each generation, variation in eye colour becomes progressively less, as the distinct colours become diluted between the two initial extremes. If this occurs, then variation in fitness will also become less and less - because it is the difference in fitness between the different eye colours that dictates this variation in fitness. Eventually, all members of the population will have the intermediate brown-blue eye colour, and there will be no variation in fitness because there is no variation in eye colour. Consequently there will be no propensity for evolutionary change. If this was the case for all features of living things, then natural selection would always grind to a halt within a short number of generations - and no directional changes in the form and function of living organisms would be possible. Without variation in form and function, there is no variation in fitness, and without variation in fitness there is no potential for evolutionary change. New mutations will introduce new features, but under blending inheritance, these features will be diluted every time the new "mutant" interbreeds with an organism with different features. As a result new mutations will be unable to spread across the population from an initial state of rarity. However, in the real world, human babies born of two parents with different eye colour normally have the eye colour of one or other parent, not an intermediate between both. This is in accord with the expectations one would derive from Mendel's experiments.

Fisher assumed that such particulate inheritance was universally true for all measurable traits, and thereby showed that if variance in form and function of living things ("phenotypic trait value") was inherited in discrete pieces, then variance in fitness would persist within populations. He therefore reasoned that selection could induce directional evolutionary change over multiple, potentially an infinite number of, generations[48]. If blending inheritance is not true, and inheritance occurs in discrete, indivisible atoms of heredity, variation in form and function is not diluted at each generation, therefore neither is variation in fitness, therefore neither is the potential for subsequent evolutionary change. Features that increase fitness can spread across populations, directional evolution can occur over many generations. The is the original concept of the gene - the atom of heredity, the physical manifestation of a discrete, indivisible difference in fitness. This difference in fitness marks out an individual possessing the gene from the rest of the population, and, crucially, cannot be diluted out of existence by the presence of other such factors. But it should be noted that the gene, in the original sense, is a theoretical necessity, resulting from the starting assumption that evolutionary change is ubiquitous, gradual, and attributable to directional selection. The gene is an assumption that dates back to a generation prior to the discovery of DNA, not an empirical observation.

Fisher's ideas revolutionized and made concrete the emerging science of population genetics. Diploid organisms, such as ourselves, possess two copies of the genetic material, one from each parent. A gene copy is known as an "allele", and different alleles may have different effects on the organisms bearing them, therefore different effects on the fitness of those organisms. The discrete nature of inheritance also permits recombination - exchange of genetic material between the copies from each parent, a process that occurs during the generation of the reproductive cells. (So for example, a person may produce a sperm or egg cell bearing, say, the eye colour allele from their father and hair colour allele from their mother, as a result of recombination). Because the combination of distinct alleles may have a positive impact on fitness, i.e. two alleles may be "greater than the sum of their parts", recombination is a significant force in evolutionary change. Evolution is a process of changing allele frequencies across populations, by virtue of their impact upon fitness. New alleles arise by mutation within the genetic material. Such mutation is, it is assumed, the ultimate source of all biological novelty. A new allele arises, by random change in the genetic material, and, if it has a beneficial impact upon fitness, increases from rarity and spreads across the population. So, the gene, in the original sense of the concept, as an atom of heredity, must be all of the following:

1. *A fitness-difference-maker*[49]. A discrete unit of difference in fitness, which distinguishes the individual with that gene from the rest of the population, and which remains an indivisible whole, in terms of its impact on fitness, during recombination.
2. *A discrete physical entity passed on from parents to offspring, rather than being acquired from the environment.* I.e. a physically discrete piece of *DNA*, which is not broken up during recombination and the generation of the reproductive material, and which is necessary and sufficient to convey the associated difference in fitness. Genes must be passed on in the germ plasm[50], i.e. the reproductive material, because characteristics that organisms acquire from their environments are not transferred to the offspring. (The rejection of acquired characteristics by Weismann contrasted with earlier theories of environment-induced inheritance by Lamarck, which do not appear to be supported by the evidence[51]).
3. *A unit of fitness-matter covariance.* The discrete physical entity in the genetic material (gene) must cause, with reliability, a particular characteristic (trait) in the organism. This trait must cause, with reliability, the individual possessing it to have the fitness difference (from the rest of the population) associated with that characteristic. At the population level, this will manifest as a gene-trait-fitness covariance. This covariance will result in genes/traits with higher fitness increasing in frequency and spreading across populations. The physical basis of the gene must cause the fitness difference directly in order for this covariance to exist, because it is the physical copy of the hereditary material (and only this) that is passed on from offspring to parents across generations.
4. *A novelty generator.* Random mutations within the genetic material must produce new gene-variants (alleles). All traits of living organisms must start out as new alleles within a single "mutant" individual. If such mutations increase fitness they will increase from rarity across the population. All new features must originate in the physical genetic material (or else they are not passed on to later generations and therefore cannot spread across the population or be subject to evolution). In other words, if biological novelty is not mappable to a gene, it is largely irrelevant to long term evolution, because it will not persist across generations.

All of the above theoretical considerations were developed well before the discovery of *DNA* by by Watson and Crick in 1953. Indeed, the search for the physical basis of the "atoms of heredity" that Fisher had proposed stimulated the search for what Schrodinger called the "aperiodic crystal"[52], the material basis of life. Without the above theoretical considerations, *DNA* might never have been discovered. Within the "modern synthesis" of biology and genetics, we are told, of course, that all variation in form, function and fitness goes through the information transfer sequence $DNA \longrightarrow RNA \longrightarrow Protein$ (and not in the opposite direction). Aspects of this conception of inheritance have proved correct. Most characteristics of living things do appear to be inherited rather than acquired from the environment. Nevertheless environmentally altered characteristics can sometimes be inherited for a small number of generations, a process associated with "epigenetic" modification of activity of the *DNA* - the environment providing cues as to which parts of the genetic material are turned on and off[53]. The basic principle of the central dogma of molecular biology, that information passes from *DNA* to protein through *RNA*, does seem to hold true, although self-replicating *RNA* and prion-proteins[54] provide examples of anomalous replication without *DNA*. There is even an argument for a limited amount of "reverse translation" (the transfer of sequence information from protein sequences to *DNA*) in the animal immune system[55], and potentially during the interaction between small, simple *RNA* and protein molecules during origin of life scenarios[56]), but the supporting evidence is extremely weak. This shows that the process is not conceptually impossible, just empirically false. Even in theory, if reverse translation were to be relevant to evolution, a living organism would need to acquire a protein from its environment (say via a symbiont), use it in such a way that it enhanced fitness, then unwind the protein into its constituent amino acid chains, and reverse translate it into its own *DNA*. Biochemically and physiologically, this would be a logistical minefield. There is no evidence that this takes place, the central dogma does appear to hold, information passes in the direction $DNA \longrightarrow RNA \longrightarrow Protein$, not the other way. Discrete pieces of *DNA* do generate discrete pieces of protein, and the sequences of both types of molecule are inherited across generations. But, crucially, do such discrete sequences correspond to discrete units of fitness difference?

The molecular biology "gene" concept is undoubtedly valid; it entirely justified to define a gene as a piece of *DNA* responsible for providing the sequence for a piece of protein. But does this gene concept match up to the evolutionary gene concept, the atom of heritable fitness difference? If the answer to this question is "yes" one would need to identify a tangible feature of a particular organism that directly impacts its fitness, and that is mappable, in a clear $1:1$ manner, on to a defined, discrete piece of *DNA*. The piece of *DNA* must be necessary and sufficient for organisms bearing it to exhibit a definable difference in survival and reproduction, compared to individuals without that piece of *DNA*. There are certain examples of this, but almost all are "loss of function" mutations (changes in the genetic material that cause a given biological characteristic to stop working effectively, as opposed to changes that produce a new characteristic). The supposedly definitive example is sickle cell anemia (taught in high school biology texts). Sickle cell disease is a hereditary blood disorder that causes red blood cells to lose their correct shape, and is mappable on to a discrete *DNA* change. Sickle cell genes are more frequent in regions affected by malaria, because the deformed red blood cells have the beneficial effect of providing resistance to the malarial parasite. People with a single sickle cell allele are partially or fully malaria resistant, a fact that is thought to have prevented the otherwise detrimental allele going extinct within malaria afflicted regions[57]. But such examples of a clear relationship between a tangible relative fitness difference and an unambiguous, discrete change in the *DNA* sequence are few and far between. It is instead generally accepted that the link between *DNA* and biological form and function is more complex. Continuously distributed traits can be explained by their being affected by multiple loci (positions in the *DNA*), which can interact with each other additively and/or multiplicatively, so called "quantitative genetics." Modern biologists speak of "genotype-phenotype mapping", the incorporation of *DNA* changes into "developmental genetic networks" and "genotype-phenotype environmental reaction norms." They do not speak in terms of tangible, specific changes in particular *DNA* sequences, causing equally tangible, specific changes in survival and reproduction. The mapping between the two is regarded as more complex and subtle, necessitating sophisticated bioinformatic analysis.

But the above discussion is emotionally charged. It is regarded as naive and ignorant to expect there to be a clear relationship between the molecular biology and evolutionary gene concepts. It is assumed that one should not expect there to be a simple and clear relationship between a specific *DNA* sequence and a specific fitness difference, except in occasional exceptionally simplistic situations. Asking "How can random changes in *DNA* cause the origin of variation in fitness?" is perceived as indicative of a lack of understanding - of the misguided expectation of simplicity in an inherently complex world, a failure to understand the timescales involved, etc. The attitude in modern biological sciences is that real relationships in biological systems are complex, have evolved over long time periods, and require sophisticated "bioinformatic" analysis - but eventually the answer will be worked out. The idea that the evolutionary gene concept simply does not warrant ontological status is a taboo, and generally regarded as religiously motivated intellectual dishonesty. A gene is a discrete piece of difference in fitness between a given individual and the population average, which parents pass on to their offspring. The difference in fitness must map on to a material entity. Because only *DNA* sequences are permanently inherited, any aspect of form, function or fitness that is not reducible to *DNA* sequences cannot be relevant to long term evolutionary trends, because it is not passed between generations. Variation in fitness must, one way or another, map directly on to tangible physical difference in the *DNA*.

In my view, it is with respect to the generation of novelty that the disconnect between the gene concept as "piece of DNA making a piece of protein" and "atom of heritable difference in fitness" is most strikingly obvious. What is the evidence that the origin of qualitatively new species during the history of life is causally reducible to random changes in DNA sequence? The ambiguity arises even in the prokaryotic world, where the alignment between the two kinds of gene concept is otherwise the closest. It is not unreasonable to imagine the gene concept as characterized above being applicable to bacteria, which share pieces of DNA ubiquitously in a universal DNA pool that has come to be known as the bacterial "pan genome"[58]. Bacteria inherit entire functional units of DNA in relatively large circular molecules called plasmids, meaning that they are able to uptake a single molecule and derive a relatively coherent functional capacity from it. But even in bacteria, the novelty generating capacity of DNA is not unquestionable. Bacteria can move towards light or a food source, can exhibit exceptional metabolic flexibility and co-operative behavior (for instance the capacity to form "microbial mats"). All such complexities of form and function, even in bacteria, came into existence, from a state of non-existence, at some point during the history of life. This is supposed to have occurred by random changes in DNA causing, by luck, coordinated physiological features, which were refined and improved upon by natural selection.

A gene must act as not only a discrete packet of inheritance, but as an instance of novelty, a new piece of heritable fitness difference. The problem is not explaining how variation is selected once it already exists, the problem is explaining how sufficient variation, appropriately tailored to what will increase fitness in the natural environment, is generated from random changes in DNA in the first place. Selection can only act upon the variation that already exists. To give a simplistic/crude metaphor, in a population of horse-like animals that form the population ancestral to giraffes, a mutant individual with a longer neck must appear, by random changes in DNA, causing random changes in protein sequence, in precisely the population and environmental context in which such a long neck would be advantageous (presumably one in which leaves on trees are nearly out of reach and the horse-like organisms must stretch to reach them, though some other just so story might also be imagined). The difficulty is not explaining how a mutant with a long neck survives and reproduces better than the rest of the population once it is already present. The difficulty is believing it arose by luck, as a consequence of random DNA changes, in exactly the environment in which this very trait could provide such a benefit. This is the problem of the origin of variation.

How can the origin of precisely the "right" variation occur without any communication between the physical environment and the hereditary material? Are long timescales and large population sizes really a sufficient explanation? The question of intermediate forms is, of course, the famous "missing link" idea, frequently voiced by creationists, and met with sneering and derision by evolutionary biologists. We are told that one should not expect to find such intermediate forms, because they have have been eliminated by natural selection. The expectation that such intermediate forms should exist reflects a fundamental misunderstanding of the evolutionary process, an inability to imagine the timescales involved, and so forth. It is tacitly assumed that in our culture, reasonable, logical people, who reject superstition in favour of ideas that can be supported by evidence, all "already know" that science already has, in principle at least, the answer to the questions of biological evolution. There may be gaps in our knowledge as to the details, there may be more that needs to be worked out via genetics, "genomics", "systems biology", the next information-age pseudonym for the field responsible for determining how life works. But the modern synthesis already has the framework for, the fundamental principles of, the answers. In this sense, within post 20th century science-centric western culture as a whole, it is tacitly assumed doubting that random changes in DNA, combined with natural selection, ultimately created the entirety of biological form, amounts to doubting the power and validity of science and reason, in favour of religion and superstition. People who believe in reason, science, logic and progress know that natural selection acts, over incomprehensibly long timescales, upon variation within populations of living organisms, and that this variation is triggered by random changes in DNA sequence. People who do not understand how this can be true are ignorant of genetics, genomics, evolutionary theory, and the statistical power of long timescales. People who actively doubt that random changes in DNA explain the origin of variation, and the closeness of organism-environment mapping, should be treated with a degree of suspicion by those who believe in science and reason. Such doubt is, it is widely assumed, probably the reflection of an intellectually dishonest, emotionally driven wish to shoehorn the Christian God (or some other supernatural "creative force") into a rational discourse. The dismissal of such intellectual dishonesty, combined with the faith that, even if there are gaps in evolutionary theory, science will one day figure the matter out, is part and parcel of being a rational person - one who believes in reason and evidence, not emotion and superstition.

This attitude is an example of what Rupert Sheldrake calls the "Science Delusion": The belief that science already has the answers in principle, leaving only the details to be worked out[59]. The disconnect between the biochemical gene concept and the culturally symbolic gene concept is well illustrated by Sheldrake's comment that:

> "*There is a vast gulf between rhetoric about the powers of genes and what they actually do. Investors in biotechnology are swept along by the metaphors, as are the readers of popular science. The problem goes right back to Weismann, who made the determinants an active agency, controlling and directing the organism's development. In effect, he endowed a special kind of matter, the germ plasm, with the properties of the soul.*

> *Genetic programs and selfish genes are similarly endowed with vital powers, including the ability to 'mould matter' and 'create form'. Thanks to the discoveries of molecular biology, we know what genes actually do. They code for the sequences of amino acids that are strung together in polypeptide chains, which then fold up into protein molecules. Also, some genes are involved in the control of protein synthesis. DNA molecules are molecules. They are not 'determinants' of particular structures, even though biologists often speak of genes 'for' particular structures or activities.....Genes are not selfish and ruthless, as if they controlled gangster homunculi. Nor are they plans or instructions for organisms. They merely code for the sequences of amino acids in protein molecules."*

Analogously, philosopher of science Paul Griffiths (whose views are more moderate than Sheldrake's) has described the concept of genetic information as a "metaphor in search of a theory"[60], noting how a software-hardware dichotomy is projected on to the biochemical reality of living things because we are so culturally familiar with such technology driven concepts - as opposed to because the scientific data demand them.

The woody vine *Boquila trifoliolata*, which grows by wrapping itself around the stem of multiple different host trees at the same time, is able to change the size, shape, colour, texture, orientation and venation pattern of its own leaves so as to match the patterning of the leaves of the host plant(s) upon which it grows. The vine is able to mimic multiple different hosts at once, causing its own leaves to blend in to the background provided by the leaves of each different plant it grows across. This blending in is so accurate that the vine's leaves are visually near indistinguishable from those of the (multiple, different) host plants - and as a result may protect the vine from herbivory[61]. The vine's phenotypic flexibility might, for argument's sake, be attributed to its own DNA sequences to and the means by which they have been moulded by natural selection during its evolutionary history. But the vine has different DNA from the host plant on which it grows, and yet it is able to adopt a form that is nearly indistinguishable from that of the host. The vine's own different leaves share, of course, the same DNA, but are able to not only detect the different host plant leaves across which they are growing, but also to mimic their form with precision. Again, different imaginable interpretations of this interesting scientific result provides an illustration of the unhelpfully polarized nature of "science-religion" discussions within our culture. Reasonable, rational people know that there is no great mystery to the vine's polymorphic mimicry, the differences arise from changes in the activation of DNA, therefore protein activity, therefore form, in time and space. The vine must be able to detect which host it is on, presumably via some hormone or emitted physiological by product by which host plants differ, and modify its own form and function according, "just" by changing "gene regulation patterns." Anybody who finds this sort of explanation unsatisfying should be viewed with disdain, because they do not adequately understand the science of evolutionary biology. Anybody who actively doubts that this is indeed the explanation probably has some sort of socio-political agenda that is not far off some variant of Christian creationism.

The juvenile cuckoo, as everyone knows, is reared in the nest of a bird from an entirely different species, it hatches earlier than the host chicks, and grows faster. Cuckoo chicks leave the nest after around 19 days, and never meet their genetic parents. Cuckoos migrate from Europe to Africa each summer, and may cross the Mediterranean and Sahara in a single flight of 1875 miles (3000km)[62]. The juvenile cuckoo flies, alone, to the adult breeding site in Africa, having never met another member of its own species. It must know where to go, through some innate disposition. The "magnetic signature" within the brains of birds (which is often touted as some sort of complete explanation for the incredible navigational accuracy of bird migration) could only tell the bird its latitude, not its precise location. But again, any deviation from the point of view that the innate migratory disposition of the cuckoo (or, of course, any other bird, or indeed any other instance of "instinctive" but complex animal behaviour) is ultimately attributable to random changes in *DNA* and natural selection over long timescales, is met with negativity, and often outright hostility. Asking questions like "How on Earth does the juvenile cuckoo make it to Africa on its own?" is, one way or another, inferred to reflect an irrational attitude that is hostile to the science that produced these (fascinating) facts in the first place.

Selection selects, it doesn't create

To return to the simple metaphor above, in a population of horse like creatures, all stretching to reach leaves in exceptionally tall trees, random changes in *DNA* sequences during the generation of the reproductive cells must cause a mutant "girraffe-horse" to occur with an exceptionally long neck. There is, we are invited to believe, no connection between the fact that the "horses" of one generation are all stretching their necks to reach high up into the trees, and the fact that the reproductive cells that they generate contain random *DNA* changes that produce longer necks: it is entirely a matter of luck. Another famous example refers to the fact that baby camels are born with callousing on their knees, which prevents burning of the skin when the animals kneel on hot sand. We are invited to believe that at some point during the history of this species, there was a population of camel like organisms that lived in a desert environment in which kneeling on hot ground was a regular occurrence, and resulted in burning of the skin that was detrimental to survival. At some point a random mutation in a *DNA* sequence during the production of the reproductive cells gave rise to a baby camel-like organism with knee callousing, which did not experience such burning. Again, crucially, there is no direct connection between the fact that knee callousing would be beneficial to survival in these ancestral camel like organisms, and the fact that DNA changes happened to produce exactly that. Explaining how a camel with knee callousing, or a horse-like creature with a longer neck, survives better and leaves more descendents once it actually exists is straightforward. The evidence from comparative anatomy for the occurrence of natural selection is insurmountable. Offspring correlate with their parents, some of these correlations lead to differences in fitness, and as a result directional evolutionary changes occur. But what, precisely, is the novelty generating process? Where does the origin of variation come from? The problem is stretching one's imagination to believe that random changes in DNA not only are undamaging to the sequence and function of existing proteins, but happen, by luck, to produce exactly the property that would benefit the organism in the natural environment. The problem is explaining the apparent coordination between the initial, pre-selection variation in fitness and the requirements of the natural environment, and reconciling this coordination with the fact that the process that generates this variation, random changes in DNA is entirely uncorrelated to the needs imposed upon organisms by the natural environment.

An evolutionary biologist reading the above paragraph would probably pose various dismissive responses: The above argument is laughably simplistic and reflects a failure to understand the timescales involved in evolution. Variation is more striking than one would imagine based on natural environments today; "camels" with knee callouses, and skin modifications all around their bodies arose in the evolutionary past, they simply did not survive. Giraffe like organisms arose with all kinds of body dimension changes - but only those with changes beneficial to survival survived. There is an inherent "observation bias" within natural selection. Obviously, today, we only see the variants that are well adapted to modern environments, because the others were eliminated during the course of evolutionary time. We do not see such extreme variation in modern populations; partial vine mimics that look only slightly like the vegetation they blend into, partial giraffes with intermediate necks that can reach some but not all of the vegetation they are stretching for (obviously these examples are intended as crude, communicative caricatures), and so forth- because the occurrence of such variation is slight, and changes within it are gradual. The expectation that we should see such variants reflects, again, a failure to understand the disconnect between short term ecological timescales and long term evolutionary timescales. The expectation that a single DNA change should "cause" such complex traits reflects ignorance of developmental genetics and the details of real biological systems. These and equivalent arguments amount to an appeal to the authority of long timescales, and a negationist dismissal of any deviation from the paradigm.

One can regard the tie between the origin of variation and random changes in DNA as a legitimate property of nature that operates over incomprehensibly long timescales, or a dogmatic and blinkered assumption prevalent within a field of biological science motivated more by dismissing creationists than explaining natural phenomena. In my view (as the reader will no doubt have gathered), the second option is closer to the truth. The molecular biologist's gene concept (as a piece of DNA that produces a piece of messenger RNA and therefore a piece of protein) is perfectly useful, and is, obviously, of extreme biochemical and biomedical importance. But the population genetics gene concept (a heritable physical entity that acts as a fitness difference-maker) persists only by virtue of a culture of faith in the future explanatory power of the scientific method. The timescales involved mean that no certain answer is possible. I would suggest to the reader that, in addition to making up their own mind about the matter, they view with suspicion anyone advocating dogmatic certainty towards this issue (wherever they fall on the continuum between Christian creationist and materialist-atheist). Most importantly, it is neither unscientific nor irrational or unreasonable to doubt the novelty generating capacity of random DNA mutations.

DNA exists, but genes, as atoms of heredity, simply do not. DNA sequences do control RNA sequences, which do control protein sequences, which in turn do, in turn, control the chemical reactions that happen in living cells. But the gene, as a physical manifestation of evolutionary fitness difference, and, particularly, as the physical manifestation of a novel instance of ordered biological fitness difference that is subsequently exposed to selection, is a socio-cultural delusion. It is necessary to consider alternative arguments for the basis of biological novelty. The reader should bear in mind that what is required of any such mechanism is that it acts as a *source* of variation in fitness, which includes the beneficial variation in which life is suited to its environment, and that any such variation is heritable, passed on to the next generation. This variation must exist prior to the operation of natural selection. So what are the possible alternatives?

Symbiosis and symbiogenesis

The word symbiosis means "the living together in permanent or prolonged association of two organisms belonging to different species"[63]. A symbiosis is a union between two (or more) distinct species with separate evolutionary histories, separate genomes, separate structures. A continuum exists in the extent to which two species within a symbiosis come to interact, the extent to which they exchange DNA, whether or not they merge permanently so as to become incapable of independent reproduction, and so forth. At one end of this continuum, there exists a competitive mutually exploitative interaction, in which separate species attack each other, extract resources from each other, perhaps cause disease or death in each other, etc. This is parasitism; when one (parasitic) organism obtains nourishment and protection at the expense of another (host) organism. Most diseases fall into this category. At the other end of the continuum is a mutually beneficial interaction in which both organisms share resources, one protects or shelters the other; mutualism. A classic (and climatically important) example of a fairly mutualistic symbiosis is a lichen - a photosynthetic cyanobacteria living inside the hyphae (cell body) of a fungus - from which it derives nutrients and to which it provides organic carbon. When two species interact, the direction this interaction takes, in terms of evolving to either mutually exploitative parasitism or mutually beneficial mutualism, depends on the extent to which the evolutionary interests of the two species align. If the survival and reproduction of each species is enhanced by the same sort of traits and environmental conditions, and the colonization of the same kinds of environments, the merger will become permanent, the *DNA* will fuse, and the two partners will come to reproduce as a single "unit of selection" (a mutualistic symbiosis). If the fitness interests of the two species do not align, one will exploit the other and the interaction will remain transient. Importantly, the physiology of a single symbiotic functional unit is frequently "greater than the sum of its parts." Symbioses can colonize environments that neither species could alone. (For instance a lichen can extract nutrients by weathering a rock surface, like a fungus, but can do so without a supply of carbon for respiration, unlike a free living fungus). The general principle that does seem to hold to a large extent is that symbiotic interactions generate physiological novelty that was not present in either of the separate species initially, and can therefore survive in environments in which neither species could alone.

The creative role of symbiosis was first emphasized by the Russian biologist Kozo-Polynsky[64]. Symbiogenesis, i.e. the generation of new types of life as a consequence of symbiotic interactions, was championed in the west by Margulis[58]. The emphasis on the creative role of symbiotic interactions during biological evolution has spawned an entire field of biological thinking, and initially controversial conclusions are now widely accepted. For instance, today, most biologists agree that the cells of eukaryotes (non-bacteria - large complex cells with the *DNA* contained in a nucleus, including animals, plants and fungi) result from the symbiotic aggregation of prokaryotes (simpler cells without a nucleus like bacteria). The classic example is that the mitochondrion (the site of cellular energy generation following the breakdown of food during respiration) is the descendent of a bacterium that became engulfed in a larger cell[65]. There are various other examples of this sort of endosymbiosis (symbiosis in which one organism lives inside another) generating fundamental eukaryotic features. The general principle is that novelty can derive from symbiotic interactions, in the sense that the two organisms complement each other physiologically, become greater than the sum of their parts. When this occurs, fusion of the *DNA*, and exchange of parts of the genome may happen. As a result, the two organisms become a single unit of selection, a single entity that passes on traits to its descendents. In this kind of permanent merger, the synergy that results from symbiotic interactions can be a direct source of the origin of heritable variation in fitness. Even though all of the *DNA* sequences, and all of the physiological properties, involved in a given symbiotic interaction were already present in one or other of the partner species before the merger began, if the interaction between the two species causes them to become greater than the sum of their parts then (i) physiological novelty can result - variation in fitness can be created, and (ii) subsequent mixing/exchange of the *DNA* will mean that this novelty can be passed on - variation in fitness will become heritable. Thus, symbiosis and symbiogenesis provide, to some extent, a conceptual alternative to random *DNA* changes alone, with respect to the origin of variation problem. Random changes in *DNA* are still, of course, necessary to generate the genome and physiology of each partner species entering in to the symbiosis, but the symbiotic whole is greater than the sum of the parts and in this sense genuine novelty can result. The requirement that the fitness interests of the two partner species are aligned amounts, of course, to saying that in terms of the repetition of their life cycles, the two species "function in similar ways." But the very fact that two fundamentally different species, with different physiologies and different organism structures, must be involved in a symbiosis, is also telling. Symbiosis, a ubiquitous feature of the living world, is a definitive illustration of the general principle of novelty resulting from the interaction between systems that are functionally similar but structurally different.

Emergence and self-organization

Emergence can be defined as a situation in which a large entity exhibits a patterned state as a result of interactions between the smaller entities of which that large entity is composed. Self-organization can be defined as a system undergoing a change from a non-patterned state to a patterned state, without any external influence acting upon that system. The idea of emergence in the simplest sense, i.e. of a whole greater than the sum of its parts, dates back to Aristotle. But the word emergence was first used by Lewes[66], who distinguished an emergent, system-level property resulting from interactions, from a "resultant", component-level property resulting from efficient causes:

> "*...every resultant is clearly traceable in its components, because these are homogeneous and commensurable. It is otherwise with emergents, when, instead of adding measurable motion to measurable motion, or things of one kind to other individuals of their kind, there is a co-operation of things of unlike kinds. The emergent is unlike its components insofar as these are incommensurable, and it cannot be reduced to their sum or their difference.*"

Philosophically, emergence can be categorized under the general label of supervenience, which roughly means "reducibility to and dependence upon lower levels":

> *"Properties of one kind, F, supervene upon those of another kind, G, when things are F by virtue of being G. Thus, a person cannot just be good, but must be good in virtue of possessing other properties, such as courage or kindness.....if one thing possesses the underlying properties and is F, then any other thing with the same underlying properties must share the resultant property F....biological properties plausibly supervene upon chemical ones; mental properties upon physical ones......One promise the notion holds out is that by its means we can understand the relation of such different layers of description without attempting a reduction of one area to the other."*[67]

In other words, if one set of properties supervene upon another, then the second set are sufficient for the first, explicable in terms of it. The key notion relating to emergence is the suggestion that just because one set of properties is necessary for another set (as, say, chemistry is necessary for biology), the second set of properties does not necessarily need to be fully explicable in terms of the first (biology is chemistry plus spatial compartmentalization and evolution, hence biology does not reduce to chemistry). If emergence is real, then supervenience does not translate into complete causal explicability. One could, of course, take the attitude that emergence is illusory and the full complexity of natural explanatory hierarchies has yet to be discovered. But in practice such an attitude tends to go hand in hand with the starting assumption that it is preferable to attempt a reduction of higher level properties to the properties of the smallest constituents of the systems in question. The concept of emergence is centered around the idea that this reduction is not necessarily justifiable *a priori.*

Various definitions of emergence directly link the process to novelty, associating the transition to a patterned state to processes that have never previously occurred. Emergence can even be defined as "contingent novelty"[68]. The application of the idea of emergence to evolutionary biology in the modern sense is summarized in Morgan's work on "Emergent evolution"[69]. This is the idea that during the history of life certain critical properties, including those properties necessary for major evolutionary transitions[10], have come into existence within systems, as a result of the rearrangement of the constituents of those systems. One might even argue that there is a direct parallel between biological emergence and Kant's notion of teleological, irreducible, natural purpose; an emergent biological property might be defined by virtue of its having a level of teleological (non-conscious) purposefulness that is not reducible to its constituent parts. This boils down to a simple idea: A system, composed of many constituent parts, can exhibit an ordered/patterned state when those constituent parts are arranged in one manner, but the same system will not exhibit pattern or order if those same constituent parts are arranged in another manner.

Although this is undoubtedly true, and undoubtedly relevant to the order (or lack thereof) of living systems during the history of life, the question is "how?" What is meant by "order", or "pattern"? If such order is genuinely novel, what distinguishes repetitive order that has always been present in the constituent parts of living systems from novel order that (literally) emerged at a defined point in time? How is this order (or lack thereof) relevant to the origin of heritable variation in fitness?

Kauffman's hypotheses concern emergent self organization at the "edge of chaos"[70]. His work adds to the basic idea (of pattern occurring in a system when its constituents are re-arranged) by being more precise about the kind of re-arrangement that will produce an ordered state. Kauffman's $N : K$ model represents a network like structure, in which the network could refer to interactions between genes, cells, even species within ecosystems. In Kauffman's model N denotes the number of nodes in the network (things being connected) and K denotes the average number of connections to other nodes per node (i.e. number of connections to other things that the average thing has). When the state of a given node is binary (0 or 1), and the state of one node influences that of those to which it is connected, Kauffman found that ordered pattern within the network occurred when the $N : K \simeq 2$. Importantly, if the $N : K$ ratio was near that conducive to order, but not quite at the appropriate value, the system became more chaotic. In other words, the transition to order occurred in close proximity to the system's most chaotic possible state- "order crystallizes at the edge of chaos." But Kauffman's $N : K$ model is difficult to extrapolate to systems in which the nodes of the network exhibit more than one possible state. In a wider sense, the relevance of emergence to novelty is vague, and descriptive rather than predictive. Why do some rearrangements produce pattern and others not? If new pattern happens in a biological system (organism), and if that new pattern is relevant to fitness, why do some organisms exhibit the pattern and others not? If all organisms exhibit the pattern, it is irrelevant to fitness and must just be hard-wired into how the form of a given species occurs. If so, where do new emergent types of form come from? Systems, including biological ones, exhibit pattern that is not reducible to efficient causes rooted in their constituent parts. But emergence as a concept is incomplete, because it fails to provide a cause for the connectivity patterns that give rise to order.

Morphogenetic fields and morphic resonance

Rupert Sheldrake has developed a (highly controversial) theory based around the hypothesis that form (i.e. shape) is inherently repetitive, and reinforces itself by access to a collective memory pool inherent in nature, without the requirement for any spatial/temporal connection or causal interaction in the conventional sense. Sheldrake calls his idea "formative causation" or "morphic resonance"[71,72]. He argues that any particular instance of an arrangement of matter (or indeed, presumably, mind) is inherently coupled, across time and space, to some form of "memory field" derived from previous instances of that same arrangement. The influence of this memory field becomes stronger the more times a given arrangement has occurred; i.e. nature is inherently repetitive, and the strength of repetition increases over time in a kind of self-referential positive feedback. For example, say an industrial chemist synthesizes a new molecule in a laboratory in one part of the world, one which does not occur in nature without human intervention. Sheldrake's theory would suggest that the fact that that molecule had been physically produced in one part of the world, would make it inherently more probable that that same molecule would form (under the same conditions), in an entirely unconnected part of the world (without communication between the chemists involved, without any physical causal connection, etc). It is, under the hypothesis of morphic resonance, as if there exists a "molecule X field", and when the actual molecule X forms, it begins to resonate with this field. This resonance becomes stronger the more times the formation has occurred - which makes it easier, as time goes on, for the constituents of molecule X to tune in to the memory field and form a physical example of the molecule (analogous to when a physical system vibrates with increasing intensity at a specific frequency, i.e. resonance). Molecules do indeed get easier to produce over time, but industrial chemists attribute this to improvements in purification techniques. Sheldrake has conducted experiments to control for this, and believes (controversially) that the results support his ideas, rather than purification artefacts (to which, it should be mentioned, most chemists would attribute the findings).

The explanatory power of morphic resonance within biology could, in theory, be highly significant. All biological species would inherently have access to a collective memory pool of their specific type of form, to which they continually referred during their development. The order of development would not need to be reducible to DNA sequences or protein sequences, they would just be aspects of a whole physical/biological system, tuning into and interacting with, the morphogenetic field. The same argument could apply to memories and thoughts; morphic resonance could provide a clear reason as to why modern science has failed to find the memory trace (the discrete, physically tangible/specific manifestation of a specific memory in the brain). Within the theory of morphic resonance, the brain would be both a generator and a receiver of consciousness. The physical basis of the brain is involved in generating thoughts and memories, but only in the same way as the physical basis of the DNA is involved in developing form during biological development - as one physical constituent of a system that tunes in to a wider field (receiving memories/development patterns by "resonating" with morphogenetic memory/development fields respectively). In Sheldrake's own words:

> *"Morphic resonance is a process whereby self-organizing systems inherit a memory from previous similar systems. In its most general formulation, morphic resonance means that the so-called laws of nature are more like habits. The hypothesis of morphic resonance also leads to a radically new interpretation of memory storage in the brain and of biological inheritance. Memory need not be stored in material traces inside brains, which are more like TV receivers than video recorders, tuning into influences from the past. And biological inheritance need not all be coded in the genes, or in epigenetic modifications of the genes; much of it depends on morphic resonance from previous members of the species. Thus each individual inherits a collective memory from past members of the species, and also contributes to the collective memory, affecting other members of the species in the future."*

Importantly, Sheldrake's ideas are testable, and the resulting are most intriguing when it comes to the psychological experiments he conducts on telepathy (which he regards as a routine means of communication in the animal kingdom) and collective memory pools. But orthodox members of the scientific community are, unsurprisingly, highly skeptical of Sheldrake's ideas.

But morphic resonance refers only to repetition, creativity is explicitly excluded from the hypothesis. Thus, a given biological entity could, hypothetically, organize the development of its form and function through morphic resonance, and different genotypes able to tune in to slightly different morphogenetic fields could exhibit different predispositions to survival and reproduction - i.e. morphic resonance could be fully compatible with the conventional notion of natural selection. For each identifiable instance of form, each trait of an organism, each physiological function, each species, etc, a morphogenetic field is postulated. The relative abundance of the real instances of form corresponding to such fields may change as a result of differing degrees of "resonance" between the fields and the actual systems. But where do the fields come from in the first place? The eukaryotic cell (as an example of a tangible system that would presumably have a corresponding morphogenetic field within Sheldrake's theory) has not always existed, but came into existence at at defined point in evolutionary time. Did the morphogenetic field come into existence at the same time? How? Did the morphogenetic field always exist? If so, why did it take so long for the physical system to come into existence? Sheldrake's theories are a coherent (hypothetical/theoretical) explanation for repetition, and do circumvent many of the problems involved in attributing the entirety of the origin of variation to random changes in DNA. But morphic resonance simply leaves out novelty.

Irreducible complexity, specified complexity and intelligent design

The controversial nature of irreducible complexity, and the view (correct or not) of most working biologists and many members of the public that it is associated with an overt socio-political religious agenda, needs no introduction. The question is whether the concept is of any value in explaining biological order or novelty. An irreducibly complex system is a system that will completely fail to function if one of its constituent parts is removed or prevented from working[73,74]; irreducible complexity refers to an inherent wholeness, an indivisibility. As Dembski puts it:

> "*A system performing a given basic function is irreducibly complex if it includes a set of well-matched, mutually interacting, non arbitrarily individuated parts such that each part in the set is indispensable to maintaining the system's basic, and therefore original, function. The set of these indispensable parts is known as the irreducible core of the system.*"

Specified complexity is directed complexity (which may or may not be irreducibly complex). In the context of Dembski's specified complexity ideas[75–77], a complex pattern is one that is highly unlikely to occur by chance, whereas a specified pattern is one that admits of a concise description (such as linguistic/syntactical representation). A specified pattern is inherently non-random, and can be explained with respect to some other pattern: "*A single letter of the alphabet is specified without being complex. A long sentence of random letters is complex without being specified. A Shakespearean sonnet is both complex and specified..*" When a system's formation/existence is necessarily attributed to an event or set of events, which are both highly improbable, and which collectively conform to a pattern that can be given independently of those events - then that system is said to exhibit specified complexity[71]. Specified complexity is argued by its proponents to provide a definitive diagnostic assay for intelligent design. Intelligent design is the view that "*certain features of the universe and of living things are best explained by an intelligent cause, not an undirected process such as natural selection.*"

The argument given by intelligent design proponents is that the explanation for any given event must fall into one of three mutually exclusive categories; regularity (an event is probable and frequent), chance (an event is improbable and rare) and design (an event is too improbable to be attributed to chance and must therefore be attributed to the conscious intent of some form of intelligence). The reasoning goes that a complex system, particularly an irreducibly complex one, is highly unlikely to occur by regularity. Further, a system that is both complex and specified is too improbable to be attributed even to chance - and a conscious intelligence (for many intelligent design researchers, an intelligence that is either analogous or explicitly identical to the Judeo-Christian God), must be invoked - in order to circumvent the inherent improbability, and to explain the direction from which the specification came. However, evolutionary biology has an additional category with which it claims to be able to explain highly improbable but specific pattern: necessity. For instance Pigliucci's comment [78] that:

> *"if a given phenomenon occurs with low probability and also conforms to a pre-specified pattern, then there are two possible conclusions: intelligent design (this concept is synonymous with human intervention) or necessity, which can be caused by a nonrandom, deterministic force such as natural selection."*

The idea of necessity dates back to Hume, but in the context of evolutionary biology amounts to saying that a biological system is a specific way because it has to be that way in order to persist. (There is a loose conceptual similarity with the anthropic principle; the universe has to be a certain way for us to exist, therefore we should not find it improbable that it is indeed that way, because we do indeed exist). Necessity attributes the features of living things to the fact that they have to be that way in order for those things to have survived. This falls back on the appeal to long timescales, large population sizes, and the implicit assumption that generating variation in form and function, including complex, coordinated traits, will happen eventually in the limits of evolutionary time.

Attributing any and all specified complexity to necessity, as evolutionary biologists do (implicitly or explicitly), is a position of faith. Faith in the scientific method and its future explanatory power, in evolutionary theory as it stands, in the view that living things are entirely explicable in terms of rationally comprehensible cause and effect relationships - but still faith. The intelligent design theorists dispute the validity of attributing specified complexity to necessity, but instead attribute the pattern to design by an intelligence. This is also a position of faith (in the existence of an intelligence capable of designing and producing systems that are inherently improbable). Intelligent design theory is vague about how one might pragmatically identify a designed feature "in the field" (regardless of the underlying probability by which such a feature might come into existence). This difficultly is particularly striking given that the intelligence responsible for this design is presumably non-human, and therefore its design specifications need not be identifiable by, nor comprehensible to, human intelligence. For instance, it is easy to identify a spade as designed, because it is known that men make spades, but what the intelligent design "God" might wish to make need not be so easily comprehensible. Additionally, the intelligent design theorists develop a "cut off" at which they deem an event too improbable to be attributable to chance, but do so without knowing the underlying "sample space" of possible events, the probability of which they are implicitly calculating; reflected by (evolutionary biologist) Nowak's comment that "*We cannot calculate the probability that an eye came about. We don't have the information to make the calculation*"[79]. The fact that calculating the probability of the origin of the eye, from the evolution of a system lacking in eyes, is indeed impossible illustrates two things. First, the intelligent design theorists are deciding that the origin of a system with specified complexity is sufficiently improbable to be attributable to an intelligence, but doing so without anywhere near complete knowledge of the underlying probabilities of such an origin event. Second, that the evolutionary biologists are attributing the same instances of specified complexity to necessity rather than design, but are doing so from exactly the same standpoint of ignorance of the underlying causes.

Crucially, neither side in the highly polarized "evolutionary biology versus intelligent design" discussion recognizes their failure to address the question of novelty. If necessity is the explanation for specified complexity, where do necessary (i.e. indispensable, foundational) instances of biological form and function come from? Long timescales, large population sizes, and selection can only explain the persistence of such features after their origin. Similarly, even if specified complexity were explicable by an intelligence that designs things, how does this intelligence work? How did God (or whoever) intervene in building the eye, the flagellum, the eukaryotic cell, and so forth? What specifically did God do and what is the evidence for, not just the need for his intervention, but the results of it? How did this intervention distinguish the designed system from systems that were left to their own devices? How exactly does the designer work, and how does this designer produce novelty?

Speculative suggestion - The role of DNA in biological evolution is not to generate novelty, but to assimilate novelty that has been generated physiologically

I suggest that a final mechanism, developed by the biologist and philosopher C.H. Waddington, is the closest to providing an explanation (albeit it an incomplete one) for the problem of the origin of variation in fitness. The environment imposes stresses on biological form, stresses to which the physiology of the organism responds (e.g. the camel kneels on hot sand, and its skin burns, because the animal's physiology is hard-wired to produce this reaction). But one way or another, the response that the whole of the rest of the organism's physiology has, to this local, specific physiological response, directed towards the environmental stress, causes, in conjunction with natural selection as it operates at the population level, the initially environmentally sensitive physiological response to become permanent and hard-wired. The essence of the idea is, intriguingly, anticipated by Nietzsche in 1883[80]:

> "*The utility of an organ does not explain its origin; on the contrary! For most of the time during which a property is forming it does not preserve the individual and is of no use to him, least of all in the struggle with external circumstances and enemies. What, after all, is 'useful'? One must*

> *ask 'useful in relation to what?' E.g. that which is useful for the long life of the individual might be unfavourable to its strength and splendour; that which preserves the individual might at the same time arrest and halt its evolution. On the other hand, a deficiency, a degeneration, can be of the highest utility in so far as it acts as a stimulant to other organs. In the same way, a state of need can be a condition of existence, in so far as it reduces an individual to that measure of expenditure which holds it together but prevents it from squandering itself. The individual itself as a struggle between parts (for food, space, etc.); its evolution tied to the victory or predominance of individual parts, to an atrophy, a 'becoming an organ' of other parts. The influence of 'external circumstances' is over-estimated by Darwin to a ridiculous extent: the essential thing in the life process is precisely the tremendous shaping, form-creating force working from within which utilizes and exploits 'external circumstances'- The new forms moulded from within are not formed with an end in view; but in the struggle of the parts a new form is not left long without being related to a partial usefulness, and then, according to its use, develops more and more completely.*"

To put the matter more succinctly, the physiological demands of an extreme environment impose changes on form, that the permanently heritable information responds to and assimilates over multiple generations.

In 1942 Waddington conducted experiments on fruit flies in which he exposed the developing embryos of the animals to an ether, which lead to a striking change (the development of an extra thorax, or body segment in the adults - a "bithorax" phenotype)[81]. At the beginning of the experiment, no adult organisms exhibited the bithorax phenotype unless they had been previously treated with the ether at the embryo stage. The phenotype was thus a physiological reaction to an extreme environmental perturbation. However, Waddington selectively bred from those adult fruit flies that exhibited the bithorax form when treated with the ether, then applied the ether treatment to their descendents, and again selected from those organisms that exhibited the bithorax phenotype in response. After a number of generations of such selective breeding for the capacity for this particular response to the environmental perturbation (i.e. for the ability to produce the bithorax phenotype in response to the ether treatment), the response became permanent. Fruit flies were born that exhibited the bithorax phenotype without being treated with the ether as embryos. The net consequence of the process was that an initially environmentally conditional response had become constitutive (permanently on, regardless of the environment). Net inheritance of "acquired characteristics" (or more precisely "initially environmentally inducible characteristics") had occurred, which Waddington labeled "genetic assimilation." Genetic assimilation does not work at the individual level within a single generation (one cannot apply the ether treatment to a single fly, watch it produce the bithorax response, then expect to reliably produce offspring with that feature). But at the population level, when an extreme environmental perturbation induces a response, and natural (or in this case artificial, experimental) selection selects for those organisms capable of exhibiting the response, over several generations, the net effect can be the making permanent of an initially environmentally sensitive feature.

The importance of genetic assimilation is regarded as uncertain within modern evolutionary theory. The basic reasoning by which the results of the experiment would be explained would involve changes in the activity of regulatory DNA sequences, including "normal" mutations in regulatory sequences that cause the relevant trait to become permanently activated. For instance, one explanation could be that the ability to produce the bithorax phenotype was present, universally within the genome of the fruit flies, but that different individuals varied at DNA sequences dictating sensitivity to such environmental stresses (i.e. the ether). Those individuals in which the bithorax phenotype was most sensitive to the environment would, by hypothesis, be most likely to produce offspring in which this sensitivity was enhanced (i.e. in which a lower environmental stimulus was necessary to induce the trait). This could happen by random DNA changes leading to the modification of the activity of other DNA sequences affecting the bithorax phenotype, e.g. changing the level of some stress factor necessary to switch on (or off) the DNA sequences necessary to produce the bithorax form. Selection for the phenotype over multiple generations lead to permanent changes in DNA sequence that changed the activity patterns, i.e. the regulation, of the DNA sequences needed for the trait. One could even potentially postulate elevated mutation rate at such regulatory regions during the period of environmental stress- perhaps because the DNA would be unwound and transcriptionally active, therefore more exposed to mutagenic substances within the cell. The details are uncertain, but genetic assimilation is nevertheless a legitimate result - in some cases, the environmental sensitivity becomes so low that the trait is essentially on all the time. However, crucially, the physiology already has to have the ability to produce the trait. Hence genetic assimilation is an explanation for the origin of variation in fitness, rather than an exhaustive explanation for biological novelty in general (the onset of whole new body plans must necessarily be attributed to symbiogenesis, emergence, or some other process).

Nevertheless, it is possible that a significant proportion of biological evolution, particularly in complex organisms, occurs along these lines - explaining why the relationship between the environment and the variation that originates is not random (i.e. because the variation is a physiological reaction to that environment). The evolutionary change is not in terms of the origin of a new feature by random mutations, but via selection for the capacity for a physiological response to an environmental stress - as a result of which this response ceases to depend upon the environment. The assumptions that one would need to accept in order to go along with this interpretation are that organisms more likely to exhibit a particular physiological response to a particular environment are more likely to produce reproductive cells in which "normal" random mutations make that response permanent. Based on the above reasoning it is possible to sketch out a formula for the origin of variation in phenotypes (form and function) and fitness during evolutionary time. This formula cannot be the entirety of the answer (because it does not address the emergence of qualitatively new body plans, and the capacity of physiologies to generate qualitatively new features):

1. In bacteria and other simple, single celled organisms, point mutation and the direct acquisition of pieces of *DNA* in plasmids, i.e. horizontal gene transfer, is sufficient to explain the origin of variation in fitness.
2. In more complex multicellular eukaryotes, random changes in *DNA* have either an insignificant impact (for instance if they occur in "junk" regions that neither encode proteins nor modify regulation of regions of the *DNA* that do), or are damaging (because they disrupt existing protein function and physiology).
3. Large complex organisms are essentially restricted to mutations that operate within a developmental body plan. The directional, selected trends within large complex organisms are confined to phenotypes that that body plan can already generate. Evolution by random changes in *DNA* sequence and directional natural selection is largely confined to changing the level of environmental sensitivity of phenotypes that are already possible within the body plan. (e.g. the animal body plan has remained the same and merely been elaborated upon, since this taxon originated).
4. A sequence for the origin of a new trait (taking the metaphorical example of the camel's knee callouses) by genetic assimilation, might be along the lines of the following:
   (a) The capacity for the trait already exists in the organism's development (camels produce callousing on their skin if it burns).
   (b) An environmental perturbation occurs (camel-like organisms encounter hot sand and burn their knees).
   (c) Some organisms exhibit a physiological response to that environmental perturbation, which benefits their survival (some camels form more callousing more easily, minimizing injury).
   (d) As a result this subset of organisms leaves more descendents in the next generation (the minimal burning of the skin translates to more foraging for food, obtaining a mate, and so forth).
   (e) These descendents include a subset in which, due to changes in regulatory *DNA* sequences during the generation of the reproductive cells, some of the second generation of camels form callouses on their knees in response to significantly less burning than did their parents.
   (f) Because the environmental perturbation (the hot sand) persists over many generations, this subset of the second generation, in which

the physiological response (knee callousing) is more easily induced, have a reproductive advantage (they are less likely to burn their legs when kneeling, because they are more likely to have skin callousing).

(g) As a result, the reduced environmental sensitivity of the trait, as well as the trait itself, increases in frequency by conventional natural selection for the environmental tolerance, which happens by virtue of relatively simple random *DNA* changes in regulatory regions, which modify the activity of pre-existing physiological functions (make skin callousing easier to initiate).

(h) By this process of incremental directional selection on the regulatory regions (rather than the trait itself), eventually the trait's environmental sensitivity approaches zero (all camels are born with knees callousing). The net result is that an environmentally conditional trait becomes hard wired. If this hard wiring modifies reproduction (say by preventing mating with those individuals without the hard-wired, environmentally insensitive trait), this may even contribute to speciation.

5. The origin of qualitatively different plans from systems with no resemblance to them remains highly problematic, and is not resolved by this suggestion. But the explanation for the origin of qualitatively new body plans is likely to be connected to both emergence and the recombination between qualitatively different systems during symbiosis. However, this is unavoidably a vague and incomplete explanation, and there is insufficient information in the fossil record to elaborate upon it unambiguously.
6. Many foundational biological features, starting with the information-homeostasis duality, through to the cell, the eukaryotic cell, various eukaryotic body plans, and so forth, are, genuinely irreducibly complex. There is a rational legitimacy to the concept of irreducible complexity that should be decoupled from its theological and socio-political connotations. But deciding whether irreducible complexity is specified, and/or whether specified complexity is indicative of design or of necessity, is an entirely arbitrary question of perspective. Specified complexity is therefore near impossible to identify in practice without an appeal to an unknown probability space. In theory, the intervention of a creative intelligence in specified complexity cannot be ruled out. But on a practical level (away from esoteric and unknowable probability spaces) distinguishing the features that would empirically invoke intelligent design is near impossible, and the nature of the designer is so poorly defined as to render the concept of a creative, biologically interventionist intelligence, essentially meaningless.

Obviously, the role of genetic assimilation in the generation of variation in fitness is a scientific issue, and therefore one that can only be discerned by real experiments, rather than by a speculative monograph such as this. The key point is that if the origin of heritable variation in fitness is due to genetic assimilation and/or to symbiotic interactions (rather than to the unfolding of efficient causes in a relatively isolated atom of heredity), then this is more consistent with the general principle upon which this manuscript is based; novelty by a symbiosis between functional similarity and structural difference. Within a successful symbiotic interaction, the physiological and evolutionary causal chains of events within one species become intertwined with those of another, their respective functional repetition is compatible. The same argument applies, albeit more loosely, to the combination between a physiology with the capacity for genetic assimilation, and an environment inducing a "assimilatable" trait over multiple generations. The systems are very different, but the functional repetition of each of them is aligned. The reduction of the (self evident) novelty generating capacity of life on earth to atoms of heredity, which are supposedly in turn reducible to discrete sequences of DNA, is demonstrably inadequate. The role of DNA, particularly in complex eukaryotes, is not to generate novelty in form and function, but to assimilate physiological novelty that arises via their interaction with the natural environment. Genetic assimilation turns transient physiological changes, which are caused by the environmental context and its interaction with the organism's physiology, into permanent, hard-wired changes. The fact that this mechanism relies on the inheritance of forms that the organism must already have the capacity to generate physiologically, is why the evolutionary trajectory of complex organisms becomes fixed into a specific "body plan." The origin of such qualitatively distinct body plans must ultimately be attributed to some form of symbiogenesis (in the widest sense of the word, up to and including a "symbiosis" between DNA or RNA and protein at the origin of life), i.e. some form of combination between qualitatively different systems.

Consciousness: The most novel property of life

Consciousness is simultaneously the most reliable, inescapable, foundational feature of the human experience, and the most inexplicable and incomprehensible. Consciousness is the window through which we look at the world, yet we don't really know what it is. At face value, we might start by defining consciousness as something along the lines of "self awareness." But this raises an immediate circularity in terms of what we exactly mean by "self" and/or by "aware." Basic dictionary definitions[78] include "*the quality or state of being aware, especially of something within oneself*", "*the state of being characterized by sensation, emotion, volition, and thought*", "*the upper level of mental life of which the person is aware as contrasted with unconscious processes.*" More sophisticated philosophical definitions[79] treat consciousness as an umbrella term covering sentience (the capability of sensing and responding to the external world), wakefulness (actually being sentient, as opposed to exhibiting the capacity to be so), self-awareness (i.e. an awareness of the self in addition to that of the external world), the exhibition of various identifiable conscious states, and the capacity to direct ones consciousness towards various objects, either in the world, or objects of intent.

The underlying themes within the various concepts of consciousness are awareness and self-referentiality (the latter of which, incidentally, raises a direct parallel with the concept of homeostasis). Consciousness is window-like, it is some kind of means by which a perception-relationship is established towards something that is not conscious. Consciousness is self-referential - one cannot be conscious unless one is somehow aware that one is a "self", an individual entity with a separate-ness from the rest of experience. But one cannot know either that other "selves" really exist, or that one's own "self" is as genuinely separate from the rest of experience as its appears. There is no way to tell that the "I" Descarte's "*cogito ergo sum*"[82] ("I am thinking, therefore I exist", or "I think therefore I am") really has any tangible separateness (one might say "I-ness") to it, because none of us can experience anything separate from our own individuality. Ultimately, none of us can determine whether there is anything else in existence from which that individuality is separate. Perhaps it would be preferable to modify Descarte's statement to "thinking happens", avoiding the issue of the impossibility of this differentiation of individuality. But "thinking happens" is not too far away from the much cruder sounding "consciousness is" (depending on which definition of "consciousness" and "thinking" that one chooses to adopt). This reflects the unavoidably foundational nature of consciousness. Once our thinking has acknowledged the existence of this foundation it is essentially impossible for us to build upon it and say something more. The inherent intractability of consciousness as a problem arises because it combines the most personal, subjective individuality and the widest, most all-encompassing wholeness. This intractability is reflected in Blackburn's[83] definition of consciousness as:

> "*Possibly the most challenging and pervasive source of problems in the whole of philosophy. Our own consciousness seems to be the most basic fact confronting us, yet it is almost impossible to say what consciousness is. Is mine like yours? Is ours like that of animals? Might machines come to have consciousness? Is it possible for there to be disembodied consciousness? Whatever complex biological and neural processes go on backstage, it is my consciousness that provides the theatre where my experiences and thoughts have their existence, where my desires are felt and where my intentions are formed. But then how am I to conceive the 'I' or self that is the spectator, or at any rate the owner of this theatre? These problems make up what is sometimes called the 'hard problem' of consciousness. One of the difficulties in thinking about consciousness is that the problems seem not to be scientific ones; Leibniz remarked that if we could construct a machine that could think and feel, and blow it up to the size of a mill and thus be able to examine its working parts as thoroughly as we pleased, we would still not find the consciousness, and drew the conclusion that consciousness resides in simple subjects, not complex ones. Even if we*

are convinced that consciousness somehow emerges from the complexity of brain functioning, we may still feel baffled about the way the emergence takes place, or why it takes place in the way it does."

Consciousness is, of course, inherently tied to the brain. We all intuitively make an instant judgment call that organisms lacking brains are not conscious, at least not in the way that we are. But even at this level of non-intellectual, raw, intuitive judgment call, we straight away find ourselves on the slippery slope of unanswerable questions. Are all organisms with brains are conscious? If a dog is conscious, is a cat? A goldfish? An insect? And so forth. We can at least say with reliability that destroying the physical basis of the brain destroys the consciousness - or at least breaks the tie between that consciousness and that body, that physical entity. Thinking about the issue of consciousness forces us to confront the issue of mind-matter dualism, because consciousness literally is the duality between our minds and our material basis, our bodies. Is mind a property of matter, reducible to the material basis of systems (like us) that exhibit mind? If matter is that which is extended in space, then how can the simple re-arrangement of pieces of matter within space give rise to something like a conscious mind? A conscious mind, at least as we experience the possession of it, is very different from an arrangement of pieces of matter within space. So what does it even mean to say that the brain evolved by natural selection (and/or emergence) within the animal taxon, as a result of mutations causing re-arrangements of the material constituents of animal bodies? How? When we analyze this statement, is the language even meaningful? On the other hand, if mind is a general property of nature, on the same (foundational) level of the conceptual hierarchy as matter, and consciousness is "just" a particular instance of mind in the same way that a chemical atom is just a particular instance of matter, then why do we seem to need brains in order to be conscious? If mind and consciousness are transcendental and beyond material, then what is it about the brain, a material system, that connects it to mind? This is the classic mind-body problem of philosophy: If mind and matter are the same, why can't we causally explain mind in terms of matter? If mind and body are fundamentally different things, how can they affect one another? The unanswerability of such questions is highly problematic as regards novelty, because the conscious mind's capacity for innovation is (surely) the most novel system within nature.

For the last generation or so, it has been frequently asserted that computers are becoming so computationally advanced that they will soon become conscious. This sort of assertion is normally connected with either an attempt to market artificial intelligence (AI) computer research, or with some form of alarmist socio-political rhetoric about the coming displacement of humanity by something approaching the HAL-9000 supercomputer[84]. The idea that it is possible to replicate the conscious mind using a computer rests on the assumption that the conscious mind is, at foundation, a logical decision making entity. By contrast, critics of the AI project have argued that much of human behaviour is based on an intuitive interactivity that cannot be formalized in logical language and that is not reducible to a logical decision making process (the failure of the AI project since its inception in the 1960's has, arguably, vindicated this skeptical view). But, for argument sake, one might suggest that even if a lot of human behaviour is an unconscious "flowing along the contours of experience"[85], consciousness itself is reducible to intentional decision making. This sort of view goes hand in hand with the idea that thoughts have a tangible manifestation in the neurology of the brain - that when we think something, brain particles are rearranged, and one way or another this rearrangement is what our thoughts "really are", that our memories are stored in the brain, have a tangible physical basis. In much the same way as the idea that evolutionary change is the result of random changes in *DNA* sequence, the idea that consciousness is reducible to the material basis of brain functioning, and that memories are stored in the brain, has become paradigmatic. And in exactly the same way as has occurred with genes and evolutionary discussions, deviation from this paradigm has come to be regarded as indicative of an irrational, anti-science, anti-reason based attitude.

Sheldrake's concept of the "science delusion" is as relevant to the relationship between brains and consciousness as it is to that between *DNA* and the form/function of living organisms. As mentioned above, Sheldrake describes the materialist-scientific worldview, unified by the tacit assumption that science, in principle, already knows the answers, leaving only the details to be worked out. Several of the paradigmatic beliefs within this "creed" relate directly to mind-body problems. Sheldrake suggests that most modern scientists take for granted certain assumptions, all of which he himself argues to be unsubstantiated. One of these is the assumption that "*All matter is unconscious. It has no inner life or subjectivity or point of view. Even human consciousness is an illusion produced by the material activities of brains*", that "*Minds are inside heads and are nothing but the activities of brains. When you look at a tree, the image of the tree you are seeing is not 'out there', where it seems to be, but inside your brain.*", that "*Memories are stored as material traces in brains and are wiped out at death*" and (most controversially) "*unexplained phenomena like telepathy are illusory.*"[86]

But even more orthodox academics, with views significantly less controversial than Sheldrake's, agree with the basic premise that modern materialist science has failed to explain consciousness in terms of any kind of causal reducibility to the material basis of the brain. For example, philosopher Thomas Nagel[87] argues that mind is a basic feature of nature that must be explained, and that materialist (i.e. efficient cause-centric) evolutionary biology has failed to provide any sort of explanation for mind or consciousness. Nagel goes further to say that human reason is another emergent phenomenon that cannot be reduced to a lower level on the explanatory hierarchy. In other words, just as consciousness is not reducible to the material basis of brain function, so reason is not necessarily reducible to consciousness, because reason allows us to tune in to a kind of underlying consistency with the world that takes us beyond just conscious awareness of it. Nagel argues that:

> "*Just as consciousness cannot be explained as a mere extension or complication of physical evolution, so reason cannot be explained as a mere extension or complication of consciousness. To explain our rationality will require something in addition to what is needed to explain our consciousness and its evidently adaptive forms, something at a different level. Reason can take us beyond the appearances because it has completely general validity, rather than merely local utility. If we have it, we recognize that it can be neither confirmed nor undermined by a theory of its evolutionary origins, nor by any other external view of itself. We cannot distance ourselves from it. That was Descartes' insight.*"

Similarly, the idea that the memory trace (the discrete material basis within the brain of a discrete memory) exists, and that brains store memories like a database, is both pervasive and without reliable empirical foundation. Sheldrake summarizes the long history of empirical attempts to locate the memory trace, for instance, Sheldrake states that:

> "*In a heroic series of experiments lasting more than thirty years, Karl Lashley (*1890 – 1958*) tried to locate specific memory traces, or 'engrams', in the brains of rats, monkeys and chimpanzees. He trained the animals in a variety of tasks ranging from simple conditioned reflexes to the solution of difficult problems. After the training, he surgically cut nerve tracts or removed portions of the brain and measured the effects on the animals' memory. To his astonishment, he found that the animals could still remember what they had learned.........Lashley started as an enthusiastic supporter of the reflex theory of memory, but was forced to abandon it.*"

and

> "*Even in invertebrates specific memory traces have proved elusive. In a series of experiments with trained octopuses, learned habits survived when various parts of the brain were removed, leading to the seemingly paradoxical conclusion the 'memory is both everywhere and nowhere in particular'. Despite these results, new generations of researchers have tried again and again to find localized memories.....[But] once again, the hypothetical memory traces proved elusive, and once more those who searched for them were forced to postulate unidentified 'storage systems' somewhere else in the brain.*"

Sheldrake concludes that "*More than a century of intensive, well-funded research has failed to pin down memory traces in brains. There may be a very simple reason for this: the hypothetical traces do not exist.*" Sheldrake's point is that modern science exhibits a stubborn determination to reject any notion of mind-matter dualism, prove that memories are stored in the brain, that conscious has emerged from the brain's material basis during evolution (and/or is completely illusory), and so forth. This assumption is built into the culture of the modern scientific-materialistic world view. When experiments produce results that are inconsistent (or at least problematic) in terms of this world view, instead of rejecting the assumption, scientists attempt another experiment in the hope that they will get the result they want. Detraction from the paradigmatic view that mind, memory and consciousness are all reducible to matter is conflated with religious superstition and therefore dismissed.

The reader should consult the work of Sheldrake, Nagel, and others (as well, of course, as those who disagree with them) for a detailed discussion of the evidence surrounding the idea that consciousness is reducible to the activity of the brain. The question as to whether mind is a property of certain arrangements of matter, that memories (and indeed thoughts) are reducible to the electrical impulses in neurons or the templating properties of brain cells, could each fill many books in themselves. For the purposes of this discussion, in terms of its focus upon novelty, I make the assumption that the following statements are self-evidently true (though the reader should note, of course, that many would dispute them):

- *Consciousness is clearly not reducible to the organization of matter within space over time.* Consciousness needs an explanatory category of its own. The conscious mind is an irreducible foundation through which we perceive anything and everything. If anything, matter and space should be reducible to consciousness, in the sense that matter and space are concepts through which we make sense of the world using our consciousness. For argument's sake, it would not conflict with consciousness as we experience it, for that consciousness to exist outside of our material bodies. I.e. we can still imagine being just as self aware as (hypothetically) some immaterial spirit-like consciousness within which we retain our minds (much like a dream). We can also imagine a hypothetical situation in which the matter and space exists but our consciousness does not, but we need our consciousness in order to undertake such an exercise within our imagination. There is no justification for the assumption that mind is reducible to matter (physicalism). Nor is there any real justification for the idea that both mind and matter should be reducible to a single entity (monism). The conscious mind and its subjectivity belongs in an explanatory category that stands alone, and does not need to be explained in terms of any other more foundational category.
- *Brain patterns correlate with patterns within consciousness but do not cause them.* Both memories and thoughts are manifested as physical interaction patterns that are necessary for consciousness to be empirically localizable to an individual body. But such patterns are not sufficient

to explain the properties of consciousness. Nor is the emergence of a brain organ during biological evolution sufficient to explain how and why consciousness exists. The brain tunes in to consciousness and memories, rather than generating them. A brain is clearly necessary for an individual human consciousness to manifest itself (at least in the "common sense" way in which we understand this manifestation), but there is no reason to presume that consciousness in a more general sense necessarily even requires a brain. (However there is perhaps justification for assuming that any consciousness that did not manifest in a human brain would be essentially incomprehensible to us). In short, mind-matter dualism is reasonable, rational, consistent with scientific investigation, and arguably even self-evidently true.

- *Consciousness manifests in living organisms as a continuum, and humans are the most conscious (known) species.* It is not that humans are conscious but that (say) chimpanzees, or dogs, or cats, mice, fish, or insects are not. The difference is one of degree, not of kind. Humans are the most conscious animal. That is, the physical basis of the human animal manifests consciousness to a greater extent than any other animal. Consciousness has generated imagination as well as reason. The conscious human mind's imagination is the most novel feature of nature, in the sense that the capacity to imagine things is not restricted to any reference point of physical possibility. The human imagination is the most novel feature of nature. This is why humanity is distinguished from other species by virtue of its capacity for technological, intellectual and artistic/subjective innovation.

The idea that human beings are the most novel species derives (as far as I am aware) from Terrence McKenna[88], who regarded human beings as being central agents in the universe's ascent into ever increasing complexity. One can, of course, attribute human novelty to the capacity for technological innovation, to language, to reason, to opposable thumbs, and so forth. All of these properties are likely necessary for the novelty generating power of consciousness. But it is not clear that this power is reducible to such properties. The human body is not very different from that of the ape. The brain has, of course, increased in size, but this does not explain the emergence of consciousness - any more than what are superficially more tangential, obscure unique human features such as (say) our capacity to stand upright and run long distances as part of hunting symbioses with dogs. The capacity for language is central because, as is widely accepted, it created a positive selection pressure for intelligence connected to social organization. But again, these are necessary events that may have been pre-requisites for human consciousness, but they do not definitively explain what consciousness actually is.

Speculative suggestion: Human consciousness is the most novel feature of nature as a result of the inherent duality that it exhibits between logical,

symbolically representable thought, and subjective, non-representable thought

Humans possess language, reason, logic, and other forms of coherent thought. But within human consciousness, this logical, linguistically representable, thought is also impacted upon by subjectivity that is not syntactically representable. When this interaction between the subjective and the objective aspects of human consciousness happens, the imagination effectively "steps outside" the rules of logical expressibility. Sometimes this results in the discovery of something that does subsequently become logically representable (a technological innovation, a new theory or artistic idea, etc). Human consciousness's innovative capacity arises because "the heart" encroaches upon "the head" in this way. Because human consciousness is the site at which the logical is continually encroached upon by the subjective, and because the combination of the two is the source of new ideas, human consciousness's capacity for logical innovation is essentially infinite. The above suggestion goes hand in hand with the assumption that we can, on some level "think what we cannot say", i.e. experience within our consciousness that which we are unable to represent coherently in logical language. This idea relates to a discussion of Wittgenstein's philosophy, and will be expanded upon in the next chapter.

Earth: the most novel known planet

The eukaryotic cell did not always exist. Nor did oxygenic photosynthesis, or animals, fungi, plants, a bacterial "pan genome", or global "biogeochemical" cycles of the major elements. The history of life on Earth is one of qualitative novelty generation. The major chemical constituents of the ocean-atmosphere system of planet Earth cycle through feedback loops connecting the land surface, the oceans and the atmosphere. Any identifiable biologically important element (say carbon, sulphur, phosphorus etc), cycles through such loops multiple times. This has lead Earth system scientists to derive the concept of a "cycling ratio"[4], the ratio between the flux of a given element within a reservoir (i.e. a standing pool of that element) and the flux entering into the reservoir from the outside. The cycling ratio measures the strength with which an element is recycled within a system, relative to the strength of the flow of that element between that system and its exterior. For carbon in Earth's oceans, the cycling ratio is about 10 : 1, which, as Volk puts it "*means that, on average, an atom of carbon will circulate among algae, zooplankton, bacteria, and bicarbonate in the photic zone about nine times, and then, after the tenth trip through the algae, it will exit....*"[89]. Thus, there is a very strong case that life has, in a general and qualitative sense, increased the strength of recycling of elements on planet Earth[4,7] to an extent dramatically distinct from those of comparable lifeless planets such as Mars or Venus. So in addition to the inherent directionality resulting from biological novelty considered in isolation, the effect of life on the physical and chemical nature of the non-living parts of planet Earth is also inherently variable. This is reflected (cautiously) by the statement by the famous Russian geochemist Vladimir Vernadsky[90] that:

> "*Life remains unalterable in its essential traits throughout all geological times, and changes only in forms. All the vital films (plankton, bottom and soil) and all the vital concentrations (littoral, sargassic and fresh water) have always existed. Their mutual relationships and the quantity of matter contained within them have changed from time to time, but these modifications could not have been large, because the energy input from the sun has been constant, or nearly so, throughout geological time, and because the distribution of energy in these vital films can only have been determined by living matter - the fundamental part, the only variable part, of the thermodynamic field of the biosphere.*"

Life keeps the Earth in a state that's good for life?

The most intriguing idea within discussions concerning the connections between life and the physical environment of planet Earth is that life itself may be a planetary phenomenon. This idea is an aspect of James Lovelock's Gaia hypothesis[91,92], which proposes climatic homeostasis by virtue of the presence of life. Lovelock's ideas differ from Vernadsky's in that Lovelock proposes active environmental optimization by life, rather than just the introduction of variability. Initially, Lovelock[93,94] argued that the fact that Mars' atmosphere was near chemical equilibrium implied that NASA's attempts to discover life there would in all probability prove to be futile. He reasoned that life was only likely to be sustained for significant periods of time when coupled to bio-geochemical cycles operating at a planetary scale. The basic empirical observation underlying this idea was that the composition of Earth's atmosphere deviated significantly from chemical equilibrium, in stark contrast to that of Mars or Venus (planets of comparable size and distance from the sun). Lovelock inferred from this that the interaction between life and Earth's climate had lead to some form of homeostatic system - that regulated the composition of Earth's atmosphere at a dynamically maintained steady state. Without such sustained, planet level homeostasis, life would never persist for an appreciable length of time. Therefore, Lovelock continues to argue, we are unlikely to find simply a few bacteria in an otherwise barren soil on some alien world. If we find life it will be detectable as planetary scale chemical disequilibrium in the atmosphere "*like that of the intake manifold of an internal combustion engine, with oxygen and combustible gases mixed, and very different from the exhausted, carbon dioxide-dominated atmospheres of Mars and Venus.*."[95]. Life, at least cellular life as we know it, cannot have originated at the planetary scale, and Lovelock's ideas implicitly invoke the general presence of dormant bacteria or spores with which planetary systems are seeded in order to get a global biogeochemical cycle going. But the uncertainties involved in such "Panspermia" ideas pale into insignificance in comparison to those involved in the origin of life. The idea that life is planetary phenomenon is entirely testable, and may turn out to be one of the great innovations in human thought.

Other aspects of Lovelock's Gaia hypothesis are less testable because they are less precisely defined. The central principle of Gaia is planetary scale homeostatic regulation- crudely, that there is some sort of way in which life keeps the Earth in a state that is good for life. Problems arise concerning the ambiguity involved in deciding exactly what is meant by "state" and "good" in this sense. Lovelock defined Gaia as some form of emergent totality arising from the interaction between life and Earth, functioning as an "adaptive control" system that modifies its context with respect to its own optimum. Gaia is:

> "*the hypothesis that the total ensemble of living organisms which constitute the biosphere can act as a single entity to regulate chemical composition, surface pH and possibly also climate. The notion of the biosphere as an active adaptive control system able to maintain the Earth in homeostasis we are calling the 'Gaia' hypothesis. Hence forward the word Gaia will be used to describe the biosphere and all of those parts of the Earth with which it actively interacts to form the hypothetical new entity with properties that could not be predicted from the sum of its parts.*"

This intriguing suggestion raised certain problems straight away. The most direct one was the scale at which homeostasis should be discernible. At local ecological scales, living things simply do not regulate their environments - they consume food from, and excrete waste products into, those environments, that is all. This and other considerations lead Volk to argue for a "by product Gaia" within which any planetary scale regulation was an emergent phenomenon derived from the local physiological by products of living organisms. But even this is poorly defined. In order to argue that a system exhibits homeostasis, that system must be both resistant to externally imposed change, and resilient in recovery from such change, and must exhibit homeostatic regulation that maintains that system at a non-random optimum state. Extreme perturbations have occurred within the history of life, most obviously the global scale snowball Earth glaciations[96], but whether life added any resistance or resilience to the relationship between Gaia and these events is almost entirely a matter of opinion.

The most challenging geochemical aspect of the Gaia hypothesis is defining what the optimum state is actually supposed to be. At ecological scales, living things quite obviously do not regulate their environments in any meaningful sense, an observation for which the abundance of evidence was recently reviewed by Tyrell[97]. Although Lovelock and supporters of the Gaia idea would argue that climatic regulation does indeed manifest at the geological scale, others would disagree. For instance, in response to the Gaia hypothesis, Ward[98] developed the Medea hypothesis - stating that complex life in particular is at least as likely to drive itself extinct as regulate its environment, and that there is a plausible case to be made that life on Earth is inherently self destructive. The discussion boils down to whether or not there exists a planetary scale, objectively definable physical/climatic state, that can reliably be said to be the optimum state for life, i.e. whether or not it can accurately be said that Gaia "*seeks a physical and chemical environment optimal for contemporary life*"[99]. Various examples of observed or inferred constancy of physical quantities (the nutrient composition or salinity of the oceans, the relationship between clouds and the fraction of incoming solar radiation reflected back to space) can be cited. Those in favour of Gaia argue that such constancy reflects planetary scale homeostasis, those against argue that it is an incidental by product that does not require any sort of active regulation. The problem is that exactly what the physical state of Earth "should" be. If climatic regulation by life is indeed occurring it is obvious that it is insufficiently well defined.

Another (more famous) problem with the Gaia hypothesis was raised by evolutionary biologists, along the lines of the fact that the Gaia hypothesis essentially ignored the information-based properties of life. Doolittle[100] noted that planetary scale feedbacks could never be encoded in the genome, and that the Earth system, because it possess no specific genome or genetic information, could not be considered alive. Dawkins argued that planetary scale regulation by the biosphere would necessitate either conscious benevolence on their part, or, at the very least, an inherent teleological purposefulness that, he argued, was incompatible with Darwinian evolution[101]. Within the orthodox evolutionary biology school represented by Dawkins, for Gaia (or any other planetary scale property) to be said to be "caused by life", it would need to be the result of natural selection. This would require the Earth to be a member of a population of other planets, all of which were reproducing, and which exhibited differing relative fitness values, i.e it would require the Earth to be a "unit of selection." This is clearly impossible because the Earth is not a replicating entity and is not contained in a population of equivalent planets with which it is competing. This issues was later clarified and resolved within the Gaia hypothesis, which now emphasizes that the Earth need not be a unit of selection, and that Gaia need not require conscious teleology. The key point is that even if, for argument sake, planetary scale regulation were caused by life, such regulation could not, by definition, be caused by natural selection. (This explains the general hostility exhibited towards Gaia by evolutionary biology, a field which is inherently uncomfortable with any feature of life not being attributable to natural selection, and which tends to conflate an emphasis on competition with theoretical rigour).

Symbiosis and the possibility of climatic regulation

Even in theory then, if planetary scale climatic homeostasis can be imagined to be caused by life, it cannot be caused by selection. If anything approaching a "Gaia" is possible, it must be attributable only to the homeostasis-based features of life, not to the information-based features. The question therefore becomes, is there anything inherent to life, which might bias the climatic systems of which life is a component part, towards homeostasis? Other than a tendency to collect together in populations and experience natural selection, due to the information replication that is part of the definition of life, does life exhibit any other general features that might, in theory, promote the existence of "Gaia"? In my own work (with my former PhD supervisors Tim Lenton and Andy Watson) I have suggested that there is such a feature: symbiosis. Symbiosis draws together species that are likely to have opposite but complementary impacts upon the climate system. Over geological timescales (in my opinion), this biases towards planetary scale homeostasis around a fixed, objective biological optimum state[102]. In order to understand what this means it is necessary to consider the daisyworld model.

Daisyworld[103] is a mathematical model that functions as a metaphor for life's interaction with the climate of planets/environments bearing life, and allows investigators to focus into mathematical form their thoughts about how life might regulate (or not) its environment. Daisyworld is an imaginary planet that derives heat from a nearby star (analogous to the Earth and the Sun). The star becomes hotter over time, so at chemical equilibrium without life, the temperature of daisyworld would also become proportionally hotter over time. The temperature of daisyworld metaphorically represents the non-living physical environment of Earth, (which, of course, the Gaia hypothesis proposes is regulated by life). Daisyworld is populated by two species of living organism - black and white daisies. The impact of the growth of these daisy species on the temperature of daisyworld metaphorically represents the impact of life on the physical state of planet Earth. Both daisy species have the same optimum temperature for growth (which is in the middle of the temperature range that the planet would exhibit just at chemical equilibrium, without any influence by life). Daisies grow fastest, and spread more across daisyworld, when their local temperature is closest to their (fixed) optimum growth temperature. The local temperature of any given daisy slightly deviates from the planet-level average. This is because the presence of the daisy species modifies the albedo (the reflected fraction of incoming radiation) of the area of ground on which it is growing. The black daisy species causes the area on which it grows to absorb more solar radiation than bare ground (because darker surfaces absorb more radiation than lighter surfaces), the white species causes the area on which it grows to reflect more radiation (because white surfaces reflect more radiation). The temperature of a given surface is proportional to the amount of radiation that that surface absorbs. As a result, on daisyworld, areas of land that are covered by black daisies are hotter, those covered by white daisies are cooler. At the planetary scale, the overall, average temperature of daisyworld increases when there are more black daisies populating the planet, and decreases when their are more white daisies.

The original daisyworld model was developed by Watson and Lovelock in response to the criticism that planetary scale homeostasis by life would necessitate teleological intention by the living species. The results of the model show that even though the amount of radiation coming in from the nearby star increased over time, the temperature of daisyworld remained relatively constant, and close to the optimum for the growth of living organisms. Because black daisies absorb more radiation and white daisies reflect more radiation, the local temperature of black daisies is always hotter than the planetary average, and the local temperature of white daisies is always colder than that average. As a result, the different impact of the two daisy species on their local temperature causes their proportions to change in such a way that planetary temperature is regulated. Suppose the star is too intense and daisyworld is hotter than the optimum temperature for growth. Then, although the local environment of all daisies will be hotter than optimum, the local temperature of the white daisy species will be slightly less hot, slightly closer to the optimum for growth. Therefore, the white daisy species will grow faster than the black daisy species, which will cool the planet. The white daisy species will continue to grow faster, spread more, until the planet is at the optimum for growth. Similarly, if the planet is too cool, the local temperature of the black species will be closer to the optimum for growth, and this species will spread across the planet, warming it. The temperature of daisyworld stays constant at the optimum temperature for growth, even though the solar energy input varies over time, because the relative proportion of the two species always adjusts naturally, so as to bring temperature back towards this optimum. Thus, using daisyworld, Watson and Lovelock were able to show that homeostasis could arise simply as a result of two species having the same optimum environmental state for growth but opposite effects on that environment. There is no need for teleology or conscious foresight, merely for simple constraints on growth and physiology.

The kind of system that the daisyworld model represents is known as "integral rein control." By "rein" it is meant that a feedback operates to "pull" a system back when that system deviates from optimum. By "integral" it is meant that a control operates over a time interval rather than instantaneously - an integral control system begins to operate when a system deviates from optimum, and continues to operate until that system no longer deviates from optimum[98]. Although daisyworld may superficially be a mere "toy" model, it has proved useful in explaining the nature of biological homeostasis. For example, Saunders[104] used daisyworld to explain why the body regulates blood sugar using two hormones (insulin and glucagon) - two reins that pull a system back when it deviates in one of two opposite directions are more homeostatically effective than one rein that can be either on or off. It is however, noteworthy, that in order for homeostasis to occur, both daisy species need to have the same optimum growth temperature. A black daisy species that warmed the planet and had a warmer growth optimum would trigger a positive feedback leading to planetary overheating, and a white species with a cold optimum would similarly trigger planetary scale cooling. Both such positive feedbacks would lead to the collapse of the system, because there would be nothing to rein in the growth of the species and its environmental influence. The key question that remains concerns the applicability of the daisyworld metaphor. Why do you need to start with two very different forms of life, with opposite environmental impacts? What real phenomenon does this metaphorical model best represent?

Speculative suggestion: If there is any feature of life that does bias the Earth towards climatic homeostasis, it is the tendency of symbiotic novelty to link together two species with similar optimum conditions for growth, but different and complementary effects on the physical environment

The natural phenomenon that the daisyworld metaphor best represents is symbiosis. A symbiosis inherently rests upon shared interest within the partner species - a symbiosis will not work unless the two species involved in it grow, survive, and reproduce under compatible conditions. But symbioses, if they are to be effective, must be between qualitatively different species. Symbioses involve costly exchange of nutrients and DNA - they will not form unless one species does something that the other derives benefit from, but is incapable of doing itself. As mentioned, the fungal component of a lichen extracts phosphorus and biochemically useful metals from a rock surface, something that the cyanobacterial partner is unable to do. But equally, the cyanobacterium is able to fix carbon, something of which the fungus is incapable. A lichen symbiosis can therefore be considered as collectively regulating a single mutually beneficial environmental quantity: the "rock nutrient : fixed carbon" ratio. Both partner species have an objective optimum value for this ratio, and both regulate the lichen's deviation from this ratio, from different directions (i.e the fungus weathers more rock and thus provides more nutrients when the ratio is too low, the cyanobacterium provides more carbon when the ratio is too high). I argue that this principle generally holds for the relationship between life and the physical environment, and is metaphorically represented by temperature in the daisyworld model. The most likely planetary scale properties that may exhibit regulation by life are not single variables, but the ratio between different biologically essential elements. Such ratios are regulated close to biological optima due to the inherent disposition of symbiosis to couple together qualitatively different physiologies in a single unit of evolution. The symbiosis can effectively take on the physiological function of whichever partner species is necessary in order for it to bring its local environment closer to optimum. Even despite the fitness cost of symbiosis, and the inherent tendency of natural selection to drive the interaction towards mutual exploitation, the physiological consequences of symbiosis tend to bias the life environment interaction towards one characterized by homeostasis. In short, symbiosis biases towards integral rein control within the life environment interaction, and integral rein control (as Watson & Lovelock showed) is, on some level, a pre-requisite for biological regulation of the natural environment.

In my own modification of the daisyworld model, I added a symbiosis between one black and one white daisy, that was able to exploit the physiology of either species (warm its environment if it is too cool, cool it if it is too hot). This symbiotic form was associated in the model with a cost of slower replication, something which natural selection is generally thought to impose cooperation upon natural populations. Additionally, I represented the fact that symbioses tend to be more physiologically robust than the partner species of which they are composed, by assuming that the symbiotic daisy was able to better tolerate deviation of local temperature from that which is optimal for growth. The result was that the symbiosis colonized the planet before the non-symbiotic species were able to grow. Overall, the range of external luminosity forcing over which life was able to regulate the temperature of daisyworld was expanded as a result of the introduction of symbiosis. In other words, the addition to the original daisyworld model of the physiological properties of symbiosis, improved the system's capacity to regulate the temperature, despite the external forcing from the star differing from that which would be optimal. In keeping with the emphasis placed by Margulis upon symbiosis in relation to Gaia, I believe something similar may have happened on the real planet Earth. If life has indeed regulated its physical environment over geologic time, it is due to the physiological effects of symbiosis, the capacity for symbiotic physiologies to produce new kinds of homeostatic systems. Symbiosis is nature's way of producing "integral rein control systems", of fusing together qualitatively different physiologies that function in similar ways.

Extreme environments and the evolution of biological cooperation

The nature of the physical environment has also impacted upon the inherent constraints on biological organization. A general pattern that is identifiable within the history of life is that of evolutionary transitions in the level of selection. Multi-level selection theory[105,106] refers, as the name suggests, to the fact that natural selection can operate at various different spatial and temporal scales. Biological information is, of course, embodied in replicators[95] (genes, cells, multicellular organisms, social groups etc). A replicator is defined as something that propagates its hereditary information, as a copy of itself, into the next generation. Evolutionary transitions involve collections of replicators that were originally autonomous aggregating together to form a single unit. (E.g. multicellar organisms are the descendents of single celled organisms that were originally capable of independent reproduction, a symbiosis forms from separate species capable of independent reproduction, etc). The aggregation process is labeled the evolution of "co-operation" or "altruism" within theoretical evolutionary biology. It is important to note that use of the word "co-operation" is, of course, not intended to suggest any human-centric conscious foresight or benevolence, but rather refers to the different levels of replicator dynamics that are involved. When a replicator (say a cell) becomes a single component within a larger group (say a multicellular organism), this generally imposes constraints on that replicator's capacity for reproduction (e.g. a cell cannot replicate in an uncontrolled manner if it is to form a tissue within a wider multicellular organism). These constraints are what is referred to when biologists speak of the evolution of co-operation. An individual replicator is "cooperative" if it restricts its individual growth/reproduction (which by definition comes at a fitness "cost" to it, in terms of lost potential reproduction) so as to adopt form and function that increases the survival and reproduction of the group of which it is a part. Within any such group, there is an inherent evolutionary pressure towards "back mutations" by which individual replicators revert to their original state of uncontrolled replication (such replicators are labeled "selfish" or "cheaters" in multi-level selection theory, though as with "cooperative" this is not intended to imply any sentience etc). When a group structure is formed from multiple entities that have the capacity for independent reproduction, two different spatial/temporal levels of replicator dynamics are relevant to the function/survival of that group. Within a group, there exists short term, individual-level replicator dynamics. For instance, within a multicellular organism, individual cells divide and reproduce over timescales significantly shorter than the lifespan of that organism. Between different groups, spatially larger and temporally longer-lived replicator dynamics also occur (multicellular organisms also themselves reproduce).

It is of fundamental importance to note that the replicator dynamics at these two different spatial/temporal scales often have different equilibrium solutions. (An equilibrium solution is one that does not change over time. In this sense it means one in which the relative proportion of different replicators within the population reaches a given value and then stays there). The existence of conflicts between the within-group and between group equilibrium solutions is the basis for multi-level selection theory. Because being a part of a group imposes a fitness cost on individual replicators (e.g. a cell within a multicellular organism must obey developmental signals that restrict its individual reproduction), then within groups, individuals that do not "co-operate" (e.g. a tumour cell that grows in an unrestricted manner regardless of the developmental signals issued to it) will, by definition, have a higher replication rate. Within a group, such "cheaters" will always replicate faster, so that once they exist it is only a question of time before they come to dominate that group. If the between group replicator dynamics are ignored, the within group equilibrium solution is complete dominance of the replicator population (of individuals) by the non-coperative type. For instance, because tumour cells grow faster than regular body cells, when the cells of the body of a multicellular organism are considered in isolation (i.e. if the replication of multicellular organisms is ignored), the only equilibrium solution is dominance of that population of cells by tumours. There are many mutations that happen in a normal body cell to give rise to a tumour, because uncontrolled replication is the ancestral state of that cell - it has the capacity for independent reproduction, and "switches off" this capacity when it becomes part of the multicellular entity. The equilibrium solution is different when the replicator dynamics of whole groups are considered. Groups of replicators composed of non-cooperative individuals have lower fitness than groups of cooperative individuals (a multicellular organism covered with tumours will die, one in which the individual cells cooperate, obeying the developmental instructions issued to them by the whole organism's body plan, will survive and reproduce). Thus, in direct contrast, if the within-group replicator dynamics are ignored, then the between-group equilibrium solution is dominance of the replicator population (of groups) by groups of cooperators.

Both within-group and between-group replicator dynamics must therefore be considered in order to understand and predict what goes on in the real world. Attempting to do so, despite the fact that the equilibrium solution of the replicator dynamics from each different level of biological organization may conflict, is the basis of multi-level selection theory. The kinds of group organization that survive and reproduce successfully involve a degree of mediation of the conflict between the within-group and between-group equilibrium solutions and/or "policing" mechanisms by which groups excise individuals that that are non-coperative (such as when a multicellular developmental system triggers programmed cell death within cancer cells). Often, the issue reduces to a "competing rates problem"[107]; the mutation towards non-cooperative individuals within a group is essentially inevitable, but if the hereditary information of that group can be passed on before this happens, then that information may still spread across the population. (For example, so long as a multicellular organism passes on its DNA before getting cancer, this DNA will still be present in the next generation, even if it contains a predisposition to degeneration to cancer). Obviously, the formation of a group is, in the first instance, an example of novelty, qualitatively different patterns of cause and effect are involved in the development of organs, a large scale body plan etc than are involved in the function of a single cell. Certain issues become apparent when considering the relevance of multi-level selection, the evolution of cooperation, and so forth, to the concept of novelty:

- *There exists a strong correlation between the relatedness between two in-*

dividual replicators (the similarity in their DNA sequences, how recently they shared a common ancestor) and the probability that those two individuals will cooperate with each other in an evolutionary sense. Cooperation between individuals is more likely to be stable in the face of natural selection if those two individuals are related, i.e. share hereditary information. This is W.D. Hamilton's famous "kin selection theory"[108]. A simple condition describes the conditions that are necessary and sufficient for two individuals to cooperate. This condition, $rb > c$, known as "Hamilton's rule", considers a single interaction between two individuals, one "altruist" individual sacrificing some aspect of its own survival/reproduction for the benefit of the wider group, and one "recipient" individual, benefiting from the other's sacrifice. r is the relatedness between the two individuals within the "genetic"/heritable material, b is the benefit in terms of increased survival/reproduction to the recipient of "altruism" (i.e. the benefit received by any given individual as a consequence of its of being part of a group of cooperating individuals - for instance the extra food, protection, resources etc that a single cell might receive as a consequence of its being part of a multicellular entity), and c is the cost to the "altruist" (i.e. the lost potential survival and reproduction that results from the individual necessarily constraining its own reproduction - e.g. the extra number of local descendents that a cell could have if it mutated into a tumour). Hamilton's rule therefore states that when the product of the relatedness between the two individuals, and the benefit to the recipient, exceeds the cost to the altruist, the evolution of altruism (costly cooperation) will occur; will increase from rarity by means of natural selection. Hamilton's rule is a simplification - there are circumstances in which close relatives compete and non-relatives co-operate, and the nature of the stability condition is affected by the population structure, and many modern evolutionary biologists argue that it is more useful to think in terms of "shared interest" between the individuals than in terms of genetic relatedness directly[109]. Nevertheless, it is a biological fact that closely cooperating entities are also often closely related. The workers of social insect colonies are all sisters, the cells of a multicellular animal all share a relatively recent common ancestor, by virtue of that organism passing through a single cell bottleneck during its development, etc. Notwithstanding the arguments above about the ultimate weaknesses of the evolutionary gene concept, one can still speak in terms of "relatedness", either in the sense of relatedness as "similarity of DNA sequence at relevant positions in the genome" or relatedness as "how recently two average individuals shared a common ancestor." All that one needs to assume in order speak in terms of relatedness is that there is a correlation between parents and offspring, and that the hereditary material is, one way or another, responsible for certain differences in fitness.

- *Intermediate levels of population mixing and dispersal of individuals between groups, favours the evolution of cooperation.* Cooperative form must spread across the population by the dynamics of individual groups, therefore some degree of mixing amongst subsets of the population is necessary for the evolution of cooperation to be stable. But excessive mixing will mean that non-coperative individuals are introduced into groups of cooperators, which will lead to the breakdown of these groups and the loss of cooperation. The evolution of cooperation is most stable in "viscous" populations[110] in which there is some dispersal, allowing cooperation to increase from rarity, but not so much dispersal that every group of cooperators is exposed to non-coperative individuals that "break" the group organization by outcompeting the cooperative individuals.

- *The arguments concerning cooperation between different individuals of the same species can be extended (with some looseness and modification) to the evolution of cooperation between species, i.e. symbiosis*[111]. In the sense of symbiotic interactions, essentially anything can act as a "replicator", a unit of selection, if it can pass sufficient information for its recurrence on to the next generation. (For instance myself and my colleagues Hywel Williams and Tim Lenton have shown how that a nutrient recycling loop, which that is an emergent property derived from the interactions of multiple different bacterial species can, within the context of an appropriate population structure, act as a unit of selection[112]). Just as there is an inherent selection pressure for non-coperative form and function within groups of like individuals, so there is an inherent selection pressure upon individuals participating in a symbiosis to evolve towards parasitism and/or mutual exploitation. Just as the avoidance of this, and the stability of cooperation is enhanced by relatedness between individuals of the same species, so increased overlap in fitness interests between two symbionts promotes the stability of the interaction. Just as restricted dispersal favours cooperation within groups of like individuals, so does restricted movement of parasites between different hosts favour the evolution of diseases to more benign "avirulent" strains. If a disease causing organism is able to rapidly kill its host then move on to a new one, it will do so and still pass on its genetic material. If such an organism finds itself in a population in which no new host is available, it will kill itself by killing its host. The general principle is that environments in which dispersal is restricted tend to lead to disease-causing parasitic organisms to evolve into more benign symbiotic organisms. (For instance the introduction of the myxoma virus to Australia failed to adequately control the rabbit population, because the virus evolved to be avirulent, probably because dispersal between individual rabbit populations was low due to the patchy population structure.[113]).

These generalizations about the evolution of co-operation can be linked together to provide a broad theme which, in my view, connects some of the major origin events that created novel biological organization during the history of planet Earth:

Speculative suggestion: Extreme environments in which population bottlenecks increase relatedness, and in which restricted dispersal makes survival a more important component of fitness than fecundity, trigger the evolution of biological cooperation. The clearest example of this process happened when global scale glaciations triggered the evolution of animals with terminal cellular differentiation. Another example may be the origin of eukaryotes.

The timing and details of the fossil record are still being determined. But two qualitative statements that continue to withstand scrutiny are that animals appear suddenly[114,115] in the fossil record, just before the start of the famous Cambrian period (from about 550 million years ago), and that the first animal like forms are found in the immediate aftermath of the most recent global scale glaciations[116] in Earth's history. A detailed exploration of all of the evidence for this statement is outside the esoteric/theoretical focus of this piece of work. It will suffice to make a two brief statements. First, during the previous "Neoproterozoic" period there is strong evidence for two, arguably three, glaciations that were global in scale. Rocks demonstrably deposited by the action of glaciers are found with magnetic properties indicating they were deposited at equatorial latitudes, and geological properties indicating that they were deposited at sea level. The implication is that global scale glaciation events occurred, immediately before the proliferation of animal life. Second, animal form emerges well after that of other large complex organisms such as algae or fungi. The obvious possibility of a causal connection between snowball Earth and the evolution of animal form is a hypothesis that I developed[117], again whilst working with my colleagues Hywel Williams and Tim Lenton. The idea is that animal form imposes a considerable fitness cost on individual cells, because in order to build an animal, cells must terminally differentiate (permanently give up their capacity for autonomous reproduction). Global scale glaciation events, I suggest, caused extreme population bottlenecks (because harsh conditions intermittently killed the majority of the population).Once a population recovers from a bottleneck, any two average members of it are more closely related (share a more recent common ancestor) than they would have been had this bottleneck not occurred (because all members of the population are, by definition, descended from the small number of survivors of the bottleneck). This increase in relatedness caused the cells of ancestral animals to be more likely to cooperate, by the process of kin selection mentioned above (if two relatives cooperate, one way or another the same "DNA for cooperation" will get passed on). This extreme degree of cooperation was necessary in order for the evolution of the "altruistic soma" - the adult body cells that permanently differentiate and thereby irreversibly give up their own capacity for independent reproduction, so that the germline (reproductive) cells are able to reproduce. This is, in my view, the key question relevant to explaining the origin of animals: Why would a cell that is originally capable of independent reproduction give up this ability, evolve into a non-reproductive differentiated cell? This question is analogous to that of why a sterile social insect such a worker ant might evolve, from an original state in which it was capable of independent reproduction, a question about which Darwin wrote:

> "*I... will confine myself to one special difficulty, which at first appeared to me insuperable, and actually fatal to my whole theory. I allude to the neuters or sterile females in insect communities: for these neuters often differ widely in instinct and in structure from both the male and fertile females, and yet from being sterile they cannot propagate their kind.*"[118].

I suggest that the problem of explaining the origin of the terminally differentiated somatic (body) cell within animal form is directly analogous to this question of the evolution of suicidally low fitness at the level of the individual. Life has solved these problems in the same way: by making the individual cells (or ants) so closely related that, as shown by Hamilton[104], the conflict ceases to be relevant, and cooperative form evolves by the inherent predisposition of close relatives to interact in a mutually beneficial way. I suggest that animals' complex body plan, with an interior and an exterior, organs, including, of course, the brain, the capacity to move, and so forth, depend on extensive differentiation at the level of the individual cell. This differentiation is more extensive in animals than any other form of life, the developmental restrictions upon the fate of the individual animal cell far exceeds that of comparable large, complex organisms such as fungi, algae, plants etc. This is why, for example, by any measure animals have more cell types, and why one cannot take a cutting from an animal any more complex than a sponge, whereas one can from a plant. In an animal the restriction upon the future reproductive potential of the individual cell is far greater than is the case in plants or fungi; the cellular fate plans of animal development impose a far higher restriction upon the individual cell.

As a result of this cost of differentiation, in terms of Hamilton's rule $rb > c$, the cost to benefit ratio $\frac{c}{b}$ is far higher for the evolution of an animal somatic cell, than for (say) a fungal cell, or a differentiated bacterial cell forming part of a microbial mat. A Hamilton's rule-like condition must nevertheless have been met in order for evolution to move from something like a sponge to something like a worm (or indeed a trilobite). Thus, animal evolution had to wait for an environment in which relatedness $r \geq \frac{b}{c}$ was also correspondingly higher. I argue that the extreme glaciations of the Neoproterozoic provided such an environment[114], by virtue of the demographic impact of the population bottlenecks that they caused. In the academic geoscience literature, I have summarized the nature of the problem concerning the evolution of the altruistic animal somatic cell as follows:

> "*Animal development requires the evolution of altruistic somatic cells with irreversible differentiation, which forgo their own long-term reproduction to facilitate that of separate reproductive cells. This contrasts with*

plants and fungi where most somatic cells are capable of de-differentiation and independent replication. A key breakthrough was probably the evolution of a cellular differentiation hierarchy between a proto-germ line retaining infinite division potential, the reproduction of which is facilitated by differentiation in other somatic cell lineages. This required multi-level selection for conflict mediation, whereby the opportunity for independent cell division by non-reproductive cells is restricted (for example, by limiting the number of somatic cell divisions) when the evolutionary interests of the cell and the organism conflict. This in turn was facilitated by sophisticated gene regulatory circuitry with conflict modifier loci, probably involving a juvenile larval stage and utilization of ancestral cellular adhesion functions. Kin selection is a mechanism by which somatic cells can evolve to not reproduce — if the soma and germ line share genetic material. This will happen if both the germ line and the soma are descended from an obligate single cell bottleneck and are thus closely genetically related. But for a single-cell bottleneck to consistently occur in an ordered manner, a tightly controlled gene regulation program is necessary in the first place — requiring altruistic somatic cells that obey developmental instructions to facilitate the bottleneck. This 'catch-22' situation is the essence of the problem involved in explaining animal origins. To explain the altruistic origins of Eumetazoa, a direct causal link with extreme glaciations has been hypothesized: in the kind of localized refugia where eukaryotic cells were likely to have survived Neoproterozoic glaciations, fluctuations in temperature, water potential and resource availability led to extreme and repeated founder effects, raising genetic relatedness and promoting kin selection for intercellular cooperation. Founder effects and restricted dispersal have been linked to the evolution of altruism both theoretically and empirically. High genetic relatedness (and low probability of successful independent reproduction) lifted a constraint on the evolution of an altruistic soma for the first time in Earth's history. This permitted the evolution, by kin selection, of an obligate separation between undifferentiated reproductive cells and a terminally differentiated soma. Once this type of development had become 'locked in', the increased degree of specialization permitted by terminal differentiation allowed the explosive exploration of form seen in Ediacaran (and Cambrian) fossils."[119]

Thus, it is possible to sketch out a hypothetical but plausible scenario for the evolution of animal life:

1. *The altruistic soma was developmentally possible but selectively disadvantageous.* Single celled, or simple multicellular ancestors of animals evolved at some point during the Neoproterozoic. They may have exhibited some kind of capacity for differentiation (perhaps when sticking together in clusters), but any such differentiation was transient, reversible and conditional upon the the physical environment - much like the differentiation observed in fungi, plants, and (to some extent) bacterial

colonies. The closest existing form of life to this sort of proto-animal like organism is possibly something like a sponge. In a sense all the cells in such a species were "germ cells", in that they exhibited the potential for independent reproduction. Because individual cells contained this potential for autonomous reproduction, formation of organs, body plans, and other complex traits requiring permanent differentiation, was significantly restricted. The inherent local selective benefit of individual replication at the level of the cell disrupted the formation of organs, or a hard-wired developmental "body plan", by causing uncontrolled cellular replication. Certain key traits needed to become fixed in the population so as to bring this replication under control. Note that this cannot be achieved by "clonal" populations of closely related cells, such as a sponge, or a collection of choanoflagellate cells. Even from an initial state in which an "altruistic soma" (i.e. a cellular form involving reproductive restraint for the benefit of a wider collective) exists, at each and every cell division event there is a non-zero probability of mutation (at the relevant positions in the DNA) back to the baseline state of uncontrolled cell division. The more cell division events occurring, the greater the probability of such mutation happening (which, of course, is why the probability of tumour formation increases with organism age). For true, cellular differentiation, a strict, controlled, single cell bottleneck must occur every generation, and each cell line must have a specific fate in a wider developmental fate plan. In other words, for true differentiation and the evolution of organs, strict controls on cell division must somehow emerge and be maintained. (In biological terminology, true "Weisman-nian development" is necessary, with a strict germline-soma distinction, and strict control on the development of somatic cells) This, I suggest, is why a collection of cells that simply aggregates cannot form an animal any more complex than a sponge (and indeed why in a sponge, aswell as in plants and fungi, the distinction between germline and soma is more facultative and more fluid). The evolution of a fixed, obligate bottleneck, through which a germline reproduces but no other cells can, literally amounts to the evolution of cellular suicide (which is why the soma is deemed "altruistic"). It is therefore impossible to imagine the altruistic soma evolving through normal evolutionary dynamics at the level of individual cells, because, by definition, it would not increase from rarity by selection within those dynamics. (Instead, I argue that exceptional environmental circumstances simply imposed the cellular bottlenecks, so that it did not need to increase from rarity, and such that the selective pressures working against it were removed). Tangible specific examples of the kind of trait that is meant by the term "altruistic soma" include the following (each of which are closely connected):

(a) A germline soma differentiation linked to obligate sexual reproduction. Many sponges reproduce sexually, but do so in an irregular manner. The reproductive cells are segregated, fuse with those of a mate organism, from which a new whole sponge grows. However, the individual body cells of sponges retain the ability to reproduce independently. This flexibility had to be lost in order for more complex, organ grade animals to evolve. Sexual reproduction had to become a hard-wired "obligate" part of the life cycle of the organism. Body cells had to evolve to give up their capacity for individual reproduction.

(b) Apoptosis (programmed cell death). Individual body cells had to be receptive to developmental signals instructing them to die at specific points in the organism's life cycle.

(c) A fixed developmental fate plan, in which every cell has a specific role and position in the final body of the organism.

2. *Snowball Earth glaciations triggered multiple, successive population bottlenecks.* Global scale glaciation events occurred. Extremes in temperature, resource availability, water potential etc killed large proportions of the population of such organisms, i.e. imposed a population bottleneck. When populations recovered (i.e. grew back to the pre-bottleneck population size), any two random individuals would have more recently shared a common ancestor, than would have two random individuals from a population that had not experienced such bottlenecks, and would be more closely related at the DNA positions responsible for controlling the cooperative traits. Repeated cellular bottlenecks were imposed upon selection pressures within populations of single cells, as opposed to evolving as a consequence of those selection pressures. Evolutionary dynamics within single celled populations favoured properties (such as cells sticking together, cells differentiating, etc) that were coordinated with the single cell bottlenecks (which happened anyway, due to regular periods of extremely harsh conditions, e.g. temperature/water potential variation, etc). The single cell bottleneck did not need to evolve from "normal" evolutionary dynamics, it was imposed by the physical environment. The selective benefit (in terms of number of subsequent descendents) resulting from uncontrolled cellular replication was far less than in more hospitable circumstances, because repeated population bottlenecks meant that populations were small, and almost no cells survived in any case. (This is what is meant by survival, rather than fecundity, becoming the dominant component of fitness).

3. *Exceptionally high genetic relatedness resulted at the positions in DNA responsible for developmentally driving terminal cellular differentiation.* The subsequent population that descended from this bottleneck exhibited higher average genetic relatedness between any two given individuals than would have been the case prior to the bottlenecks. As a consequence there occurred stronger kin selection for the cooperative traits above. I.e. particular types of form and function, such as differentiation favouring reproduction of a proto-germline but damaging the reproductive potential of the differentiated cell line itself, were favoured - because the

cooperative cells were exceptionally highly related to the beneficiaries of the cooperation. Without the repeated bottlenecks, average relatedness between cells, at the relevant positions in the DNA would not have been high enough to favour such cooperative properties.

4. *Repetition of strong kin selection and group viability over overwhelmingly long geologic timescales.* Repeated population bottlenecks raised genetic relatedness, therefore kin selection for cellular cooperation, to an unprecedented extent. Cellular clusters containing revertant mutants that de-differentiated (replicated at the level of the individual cell in an uncontrolled manner) may have gone extinct due to the lack of cooperative form (e.g. resource storage) that such clusters would have exhibited. By contrast, differentiated cellular clusters may have been less likely to go extinct, because their survival depended only upon that of a single germ cell line (which could have been preferentially segregated from the environment by the differentiated form, e.g. by an ectoderm-endoderm tissue level differentiation). Restricted dispersal associated with the glacial events limited the spread of non-coperative individual cells between clusters. In the terminology of evolutionary biology, cellular differentiation was promoted by kin selection, "group viability selection" (groups of cooperating cells survived better), and restricted dispersal.
5. *Fixation of the altruistic soma, permitting the subsequent explosive diversification of animal form.* As a consequence, a germline-soma distinction, apoptosis, sexual reproduction, and a fixed body fate plan became established, hard wiring aspects of the form of a multi-cellular animals, such that these properties were no longer environmentally contingent. Once in place, apoptosis, an obligate single cell bottleneck during the life cycle, and various other "policing mechanisms" directly eliminated single cells that disrupted organism level development, stabilizing animal form. (For example, once a developmental system was in place that "attacked" tumour cells exhibiting uncontrolled differentiation by something loosely analagous to an immune response, animal form was to some extent "locked in", despite the inherent selective benefit at the cell level for tumour formation). The key argument here is that the extreme environment gave rise to a period in which selection against cellular cooperation (i.e. selection for uncontrolled cellular replication) was relaxed, due to the high genetic relatedness, and priority of survival over fecundity. This allowed animal form to become established, probably leading to a transition from something similar to a sponge to something similar to a worm. The internal evolutionary stability of terminal (irreversible) cellular differentiation was established at the organism (as opposed to cellular) level, permitting the explosive diversification of animal form observed in the subsequent fossil record, i.e. the famous Cambrian explosion.

This scenario, as with the issues discussed above, inherently involves novelty by the combination of two distinct systems: An extreme climatic situation that imposed population bottlenecks, to a severe extent that was repeated over geologic timescales, and a developing population of single cell heterotrophic eukaryotic organisms, with the capacity to build an animal already within their *DNA*. The environment created a unique evolutionary context, high relatedness and strong Hamiltonian kin selection, by which these organisms were able to overcome the evolutionary constraints imposed upon them as a result of their being composed of individual cells with a disposition to compete with each other.

Even well before the origin of animals, a similar kind of process may have been involved in the origin of eukaryotes (organisms with large cells in which the *DNA* is contained in nuclei), however the problem is less well formulated and the evidence is more coarse. There are many subtleties and constraints concerning the evolution of eukaryotes, which render the issue largely outside the focus of this piece of work. But the key point is that (once again) the event necessarily originated from a combination of qualitatively different species. As well as a nucleus, the eukaryotic cell definitively possesses a "cytoskeleton" (a network of polymers within the cell body by which it can move and control the cell's internal constituents), and endosymbionts, such as a mitochondria, within which the transport of electrons derived from the breakdown of food substances produces energy. Endosymbionts are descended from free-living bacteria, so in order to make a eukaryote a larger cell must capture a smaller one- something which is incredibly rare in the bacterial world. A catch 22 situation arises due to the fact that a cytoskeleton is extremely energetically demanding, but the kind of energetic turnover per cell that characterizes modern eukaryotes can probably only be supported if multiple mitochondria are present within the cell and carrying out electron transport[120], the engulfing of which probably requires a cytoskeleton. I.e. the kind of apparatus necessary to capture bacteria and turn them into symbionts can only be energetically supported by the presence of symbionts. I have speculated that the initial step involved a similar kind of restriction on populations imposed by glaciation events (another set of which may have happened prior to the earliest appearance of eukaryotes in the fossil record). I suggest that extreme restrictions on dispersal, by virtue of life being forced into small habitable refugia by the extreme glaciation events, somehow lead to a large cell (the host) incorporating a bacterial endosymbiont (the bacteria), perhaps even literally as a result of being frozen into a tiny bubble of liquid water within a surrounding area of ice[121]. This idea is highly speculative, because the timing of the emergence of eukaryotes in the fossil record is highly uncertain, as is that of the global scale glaciations that may have preceded this emergence. But the key point is that natural selection and/or the inherent constraints on biological form, imposes a barrier to a given instance of biological complexity getting started (in this case an energetically demanding cytoskeleton that allows a system to take up symbionts). But then an extreme environment serendipitously causes the instance of complexity to happen despite these constraints, by changing the result of natural selection's cost/benefit analysis, and by forcing qualitatively different systems together.

Conclusion

Thus, in summary, I have suggested that the history of life supports the premise that extreme instances of novelty result primarily from the combination of inherently different systems, rather than the unfolding of the effects of causally autonomous atoms of heredity. I have suggested that symbiosis and genetic assimilation are far more likely explanations for biological novelty in a generic sense than are random changes in *DNA*. Such random changes are a necessary feature of evolutionary change, but (even given intriguing theoretical possibilities such as the idea that more transcriptionally active regions of the *DNA* are more likely to mutate, meaning that those regions that are important to fitness inherently evolve faster[122]), random *DNA* mutations are inadequate to explain the origin of complex, coordinated form. The origin of variation, the onset of precisely the particular variation in fitness that will benefit an organism in a given environment, cannot be due solely to *DNA* changes lower down the hierarchy of biological organization. Nor can it be dismissed as "necessity."

One way out might be the suggestion that most evolutionary change, particularly in complex organisms, follows something along the lines of Waddington's genetic assimilation mechanism. The role of *DNA* is not to generate novelty, but to assimilate the novelty that is already present in the physiological capacity of an organism, to make permanent an initially environmentally conditional change. This restriction to the kind of variation already present within the body plan raises the obvious question of where the body plans themselves come from. There is clearly a good deal of validity to the idea of an inherent restriction upon the laws of form, underlying the pattern of convergent evolution[123]. But the idea that such laws of form are somehow fixed defines biological novelty out of existence whilst ignoring the question of how a given body plan, a given organizational feature, actually came to manifest itself in the real world. Part of the answer is symbiotic recombination of separate systems leading to emergent properties. But this still raises the question of where the systems that symbiotically recombine come from, leading to an infinite regress, perhaps all the way back through the origin of life, to the formation of chemical elements in stars by the forcing together of smaller constituent elements. Such "symbiotic" forcing together is probably the single strongest driver of biological diversity, and indeed novelty in a wider sense.

If there is any property of life that causes biology to bias towards a Gaia, to promote a situation in which "life keeps the Earth in a state that is good for life" it is the tendency for symbiosis to stabilize integral rein control structures by fusing together qualitatively different physiologies. But there is more to biological novelty as it manifests on Earth than (only) the recombination of qualitatively different systems of biological organization. There is also the combination of biological systems with extreme environments that modify the prevailing evolutionary pressures, the conflicts and constraints acting upon the individual replicators of which biological groups are composed. The definitive example of this, I suggest, is the most powerful instance of operation of W.D. Hamilton's kin selection in the history of planet Earth- induced by the global scale Neoproterozoic glaciations. The high genetic relatedness resulting from population bottlenecks caused by these global scale glaciations lifted a selection-driven constraint on the evolution of permanent altruistic cells, triggering the evolution of animals. The general principle is one of the combination between a population with a genetic predisposition that is not manifested, due to intrinsic selection-driven constraints, and an exceptional physical environment that renders such constraints irrelevant, because survival, and not fecundity, becomes the dominant component of fitness, and survival necessitates cooperation, particularly between genetic relatives. This, I speculate, may be the key to the evolution of complex life in a general sense.

Notes

1. Lovelock, J. *Gaia: A new look at life on Earth.* (Oxford, 1979).
2. Lovelock,J. *The Ages of Gaia: A biography of our living Earth.* (Norton, 1988).
3. Knoll, A. *Life on a young planet: The first three billion years of evolution on Earth.* (Princeton, 2003.
4. Volk,T. *Gaia's Body: Towards a physiology of Earth.* (MIT, 2003).
5. Schwartzmann, D. *Life, temperature and the Earth.* (Columbia, 1999).
6. Canfield, D. *Oxygen: A four billion year history.* (Princeton, 2014).
7. Lenton, T. & Watson, A. *Revolutions that made the Earth.* (Oxford, 2011).
8. Kasting, J. *How to find a habitable planet.* (Princeton, 2010).
9. Crick, F. *Life itself: It's origin and nature.* (Futura, 1982).
10. Maynard-Smith, J. & Szathmary, E. *The major transitions in evolution.* (Oxford,1995).
11. Morowitz, H. *Beginnings of cellular life: Metabolism recapitulates biogenesis.* (Yale, 1984).
12. Morowitz, H. *The emergence of everything: How the world became complex.* (Oxford , 2004).
13. Waddington, C. *The nature of life.* (George, Allen & Unwin, 1961).
14. de Duve, C. *Vital Dust: The origin and evolution of life on Earth.* (Basic Books, 1995).
15. de Duve. *Life Evolving: Molecules, mind and meaning.* (Oxford, 2002).
16. Margulis, L. & Sagan, D. *Microcosmos: Four billion years of microbial evolution.* (University of California Press, 1986).
17. Margulis, L. & Sagan, D. *What is life?* (University of California Press, 1995).
18. Thain, M. & Hickman, M. *The Penguin dictionary of biology.* (Penguin, 2004).
19. Berg, J, Tymoczko, J, Gatto, G, & Stryer, L. *Biochemistry: 8th edition.* (Freeman. 2015).
20. Watson J.D. & Crick F.H. *A Structure for Deoxyribose Nucleic Acid.* Nature 171 (4356): 737–738. 1953.
21. Merriam-Webster online dictionary (Encyclopedia Britannica). http://www.merriam-webster.com/dictionary/information
22. Shannon, C. E. *A Mathematical Theory of Communication.* Bell Systems Technical Journal 27 (3): 379–423. 1948.
23. Schulze-Makuch, D. & Irwin, L. *Life in the universe: Expectations and constraints.*

(Springer, 2006).
24. de Duve. *Ibid.* ref. 15.
25. Kant, I. *Critique of judgement.* (Hackett Classics. 1987. (First published 1790)).
26. McLaughlin, P. *Kant's critique of Teleology in Biological explanation: Antinomy and Teleology.* Studies in the History of Philosophy 16, 1990.
27. Maynard-Smith & Szathmary. *Ibid.* ref. 10.
28. Crick F.H. *The origin of the genetic code.* J Mol Biol 38 (3): 367–379. 1968.
29. Orgel L.E. *Evolution of the genetic apparatus.* J Mol Biol 38 (3): 381–393. 1968.
30. Gilbert, W. *The RNA World.* Nature 319 (6055): 618. 1986.
31. Robertson M.P., Joyce G.F. *The origins of the RNA world.* Cold Spring Harb Perspect Biol 4 (5). 2012.
32. Cech, T. 2000. *The ribosome is a riobozyme.* Science 289.878 – 879. 2000.
33. Spiegelman, S. *Extracellular evolution of replicating molecules.* In *The Neuro Sciences: A Second study program*, pp 927 – 945 Schmidtt, F.O. ed. Rockefeller University Press. 1970.
34. Eigen, M. *Self organisation of matter and the evolution of biological macromolecules.* Naturwissenschaften 64.541 – 565.1971.
35. Maynard-Smith & Szathmary. *Ibid.* ref. 10.
36. Berg et al. Ibid ref. 19.
37. Miller, S.L.& Urey, H.C. *Organic Compound Synthesis on the Primitive Earth.* Science 130 (3370): 245–51. 1959.
38. Dale, T. *Protein and nucleic acid together: A mechanism for the emergence of biological selection.* Journal of Theoretical Biology 240.337 – 342. 2006.
39. Watson & Crick. Ibid. ref. 20.
40. Dill, K.A. & McCallum, J.L. *The Protein folding problem:* 50 *years on.* Science 338(6110) 1042 – 1046.2012.
41. Wolpert, L & Sheldrake, R. *What can DNA tell us? Place your bets now.* New Scientist, 8th July 2009.
42. Dill. Ibid. ref. 40.
43. Fisher, R.A. *The correlation between relatives on the supposition of Mendelian inheritance.* Trans. Royal. Soc. Edinburgh 52.399 – 433. 1918.
44. Fisher, R.A.. *The Genetical Theory of Natural Selection.* (Oxford, 1930).
45. Mendel, G. *Experiments in plant hybridization.* (Cosimo Classics. 2008, (First published 1865)).
46. Weismann, A. 1893. *The germ plasm: A theory of heredity.* (Electronic scholarly publishing http://www.esp.org/books/weismann/germ-plasm/facsimile/).
47. Charles Darwin. *On the origin of species by means of natural selection: or the Preservation of the Favoured Races in the Struggle for Life.* (Murray, 1857).
48. Fisher. Ibid. ref 43.
49. Sterelny, K. & Griffiths, P. *Sex and Death: An Introduction to the Philosophy of Biology.* (Chicago, 1999).
50. Weismann. Ibid ref. 46.
51. Mark Ridley (Ed). *Evolution* (3rd Edition). Wiley-Blackwell. 2003.
52. Schrondinger, E. *What is life? The physical aspect of the living cell.* (Cambridge,1944).
53. Jablonka, E. & Lamb, M. *Evolution in four dimensions: Genetic, epigenetic, behavioural and symbolic variation in the history of life.* (Bradford, 2014).
54. Prusiner, S.B. *Prions.* Proceedings of the National Academy of Sciences of the United States of America 95 (23): 13363. 1998.
55. Cook, N.D. *The case for reverse translation.* Journal of theoretical biology 64(1). 113 – 135. 1977.
56. Nashimoto, M. *The RNA/Protein Symmetry hypothesis: Experimental support for reverse translation of primitve proteins.* Journal of theoretical biology 290(2). 181 – 187. 2001.
57. Allison, A.C. *Genetic control of resistance to human malaria.* Current Opinion in Immunology 21 (5): 499–505. 2009.
58. Mira, A. Martin-Cuadrado, A.B., D'Auria, G. & Rodriguez-Valera, F. *The bacterial*

pan-genome: A new paradigm in microbiology. Int. Microbiol. 13. 45 – 57. 2010.
59. Sheldrake, R. *The Science delusion.* (Coronet, 2012).
60. Griffths, P. *Genetic information: A metaphor in search of a theory.* Philosophy of Science 68. 394 – 412. 2001.
61. Gianoli, E, & Carrasco-Ura, F. *Leaf mimicry in a climbing plant protects against Herbivory.* Current Biology 24(9). 984 – 987. 2014.
62. http://www.rspb.org.uk/discoverandenjoynature/migration/stories/cuckoos.aspx (Website of the Royal Society for the Protection of Birds, UK).
63. Thain & Hickmann. Ibid. ref 18.
64. Kozo-Polyanski, B. *Symbiogenesis: A new principle of evolution.* (Harvard, 2010).
65. Margulis, L. *Symbiosis in cell evolution.* (Freeman, 1981).
66. Lewes, G. H. *Problems of Life and Mind.* London: Trübner. 1875.
67. Blackburn. S. *The Oxford dictionary of Philosophy.* (Oxford, 2008).
68. Durham emergence Project, Durham University, UK https://www.dur.ac.uk/resources/emergence/DurhamEmergenceProjectDescription.pdf
69. Morgan, C. *Emergent Evolution.* (Henry Holt, 1923).
70. Kauffman, S. *The Origins of Order: Self Organization and Selection in Evolution.* (Oxford, 1993).
71. Sheldrake, R. *A new science of life: The hypothesis of morphic resonance.* (Park Street Press, 1995).
72. Sheldrake, R. *The presence of the past: Morphic resonance and the habits of nature.* (Park Street Press, 2012).
73. Behe, M. *Darwin's black box: The biochemical challenge to evolution.* (Free Press, 2006).
74. Behe, M. *The edge of evolution: The search for the limits of Darwinism.* (Free press, 2007).
75. Dembski, W. *The design inference: Eliminating chance through small probabilities.* (Cambridge, 1998).
76. Dembski, W. *Intelligent Design: The Bridge between science and theology.* (Intervarsity Press. 1999).
77. William Dembski. *No Free Lunch: Why specified complexity cannot be purchased without intelligence.* (Rowman & Littlefield, 2001).
78. Pigliucci, M. *Chance, necessity, and the war against science.* BioScience 50: 79–72. 2000.
79. Nowak, M. 2005 Time Magazine, 15 August 2005.
80. Nietzche, F. *The will to power.* (Vintage Books 1967. (First published1901)).
81. Waddington, C.H. *Canalization of development and the inheritance of acquired characters.* Nature, 150, 563-565. 1942.
82. Descartes, R. *Specimina philosophiae.* Elsevier. 1644.
83. Blackburn. Ibid. ref. 67.
84. Clark, A. 2001*: A space odyssey.* (Hutchinson, 1968).
85. Dreyfus, H. *What Computers Can't Do: The limits of artificial intelligence.* (HarperCollins, 1978).
86. Sheldrake. Ibid. ref. 71.
87. Nagel, T. *Mind and Cosmos: Why the Materialist Neo-Darwinian Conception of Nature is Almost Certainly False.* (Oxford, 2012).
88. McKenna, T. & McKenna, D. *The invisible landscape: Mind, Hallucinogens and the I Ching.* (HarperOne, 1994).
89. Volk, Ibid. ref. 4.
90. Vernadsky, V. *The Biosphere.* (Synergetic Press. 1986, (First published 1926)).
91. Lovelock. Ibid. refs 1, 2.
92. Lovelock, J.E. *Gaia as seen through the atmosphere.* Atmospheric Environment. 6. 579 – 580. 1972.
93. Lovelock, J.E. & Margulis, L.M. *Atmospheric homeostasis by and for the biosphere: the gaia hypothesis.* Tellus XXVI 1 – 2. 2 – 10. 1973.
94. Lovelock, J.E. *Hands up for the Gaia hypothesis.* Nature 3(44). 100 – 102. 1990.
95. Lovelock & Margulis. Ibid. ref. 93.

96. Kirschvink, J.L. *Late Proterozoic Low-Latitude Global Glaciation: the Snowball Earth.* The Proterozoic Biosphere: A Multidisciplinary Study. J.W. Schopf & C.Klein. Cambridge University Press. pp51 – 52 . 1992.
97. Tyrrell, T. *On Gaia: A critical investigation into the relationship between life and Earth.* (Princeton, 2013).
98. Ward, P. *The Medea Hypothesis: Is life on Earth ultimately self-destructive?* (Princeton, 2015).
99. Lovelock, J. *The Vanishing Face of Gaia.* Basic Books. 2009.
100. Doolittle, W. F. *Is Nature Really Motherly?* Coevolution Quarterly 29: 58–63. 1981.
101. Dawkins, R. *The extended phenotype.* Oxford University Press. 1982.
102. Boyle, R.A., Lenton, T.M. & Watson, A.J. *Symbiotic physiology promotes homeostasis in daisyworld.* Journal of theoretical biology 274.170 – 182. 2011.
103. Watson, A.J. & Lovelock, J.E. *Biological homeostasis of the global environment: the parable of Daisyworld.* Tellus B (International Meteorological Institute) 35 (4): 286–9. 1983.
104. Saunders, P.T., Koeslag, J.H. & Wessels,J.A. *Integral rein control in physiology.* J. Theor.Biol.194,163–173. 1998.
105. Michod, R. *Darwinian Dynamics: Evolutionary transitions in fitness and individuality.* (Princeton, 2000).
106. Wilson, D.S. & Sober, E. *Reintroducing group selection to the human behavioral sciences.* Behavioral and Brain Sciences 17 (4): 585–654. 1994.
107. Goodnight, C.J. *Multi-level selection: The evolution of cooperation in non-kin groups.* Population Ecology 47:3–12. 2005.
108. Hamilton, W. D. *The evolution of altruistic behavior.* American Naturalist 97 (896): 354–356. 1963.
109. Nowak, M.A. *Five rules for the evolution of cooperation.* Science. 314(5805) 1560-1563. 2006.
110. Van Baalen, M. & Rand, D.A. *The unit of selection in viscous populations and the evolution of altruism.* Journal of theoretical biology 193(4):631-648. 1998.
111. Frank, S.A. *Models of symbiosis.* American Naturalist, 150,S1. *S*80 – *s*99. 1997.
112. Boyle, R.A., Williams, H.T.P & Lenton, T.M. *Natural selection for costly nutrient recycling in simulated microbial metacommunities.* Journal of theoretical biology 312. 1 – 12. 2012.
113. Sober, E. & Wilson, D.S. *Unto others: The evolution and psychology of unselfish behaviour.* (Harvard, 1998).
114. Hofmann, H.J.; Narbonne, G.M.; Aitken, J.D. *Ediacaran remains from intertillite beds in northwestern Canada.* Geology 18 (12): 1199–1202. 1990.
115. Waggoner, B. *Interpreting the Earliest Metazoan Fossils: What Can We Learn?.* Integrative and Comparative Biology 38 (6): 975–982. 1998.
116. Hoffman, P.F. & Schragg., D.P. *Snowball Earth: Testing the limits of global change.* Terra Nova. 14(3). 129 – 155. 2002.
117. Boyle, R.A., Lenton, T.M. & Williams H.T.P. *Neoproterozoic "snowball Earth" glaciations and the evolution of altruism.* Geobiology 5(4). 337- 349. 2007.
118. Darwin. Ibid. ref. 47.
119. Lenton, T.M., Boyle, R.A., Poulton, S.W., Shields-Zhou, G.A. & Butterfield, N.J. *Co-evolution of eukaryotes and ocean oxygenation in the Neoproterozoic era.* Nature Geoscience 7 (4), 257-265. 2014.
120. Lane, N. *Mitochondria and the meaning of life.* (Oxford, 2005).
121. Boyle, R.A. *The problem of eukaryotic origins in relation to the early/mid Proterozoic Earth System.* In *Revolutions in the Early Proterozoic: Tracking Geochemical and Geobiological Change* (Edited by S. Poulton & D. Johnston D). Springer. In Press.
122. Williams, R. & da Silva, J. *The chemistry of evolution: The development of our ecosystem.* (Elsevier, 2006).
123. Conway Morris, S. *Life's solution: Inevitable humans in a lonely universe.* Cambridge University Press. 2003.

NOVELTY, MEANINGFUL LANGUAGE, SUBJECTIVITY AND TIME: A COMPARATIVE READING OF SELECTED WORKS BY LUDWIG WITTGENSTEIN AND MARTIN HEIDEGGER

Ludwig Wittgenstein and Martin Heidegger were the two most important philosophers of the 20th century, primarily because their respective works prompted a complete re-evaluation of the scope and purpose of both academic philosophy and any other form of disengaged contemplation[1]. This significance is reflected by Searle's statement that:

> "*The philosophical tradition that goes from Descartes to Husserl, and indeed a large part of the philosophical tradition that goes back to Plato, involves a search for foundations: metaphysically certain foundations of knowledge, foundations of language and meaning, foundations of mathematics, foundations of morality, etc.... Now, in the twentieth century, mostly under the influence of Wittgenstein and Heidegger, we have come to believe that this general search for these sorts of foundations is misguided.*"[2]

Wittgenstein's early work powerfully suggests that there are no true philosophical problems, but rather that the meaningless nonsense which arises from the application of a language to a context outside that for which it is designed, often exhibits a "philosophical" appearance. Wittgenstein argued that such nonsense should be swept away in favour of meditative, reverent, but absolutely unavoidable, silence. In stark contrast, for Heidegger the combination of a question's an intuitive appeal and its linguistic ambiguity was exactly what made it philosophically important. Wittgenstein focused on the prerequisites for, and implications of, genuinely meaningful language[3], exploring the premise that the intrinsic structure of a meaningful proposition must exhibit a homology to the intrinsic structure of the fact in the world to which that proposition refers. The generic nature of this homology is encapsulated in a "general propositional form" that necessarily (and to a large extent unsurprisingly) is the basis of all proper use of language. From a completely different kind of initial standpoint, Heidegger proposed that an interactive, inescapable, irreducible notion of "Being-in-the-world" constitutes a definitive (and academically neglected) core of what it means to be authentically human[4]. Heidegger was less concerned with defining in precise linguistic/semantic terms exactly what this was supposed to mean. The nature of human being in the world was necessarily ambiguous and could only be hinted in language at by semantic circularity. A comparison between Wittgenstein and Heidegger encompasses the whole spectrum of 20th century academic philosophy, because the different perspectives of these two men mark the opposite ends of a continuum between two schools of thought - each with irreconcilably different attitudes to the relationships connecting language, thought, meaning, and subjectivity. Such a comparison is also, I argue, of direct relevance to novelty.

The criteria by which meaningfulness is assigned (or not) to statements in language are foundational within the approach to any philosophical question. Insisting on the precise meaning of words before beginning a discussion can cause many, if not most, (apparent) philosophical questions to cease to be stateable, let alone answerable. For example, the question "What is the meaning of existence?" implicitly requires (if it is to be answerable) some set of criteria by which we can assign "meaning." If "existence" is to be assessed according to these meaning-criteria, then the criteria must have some form of consistency that is separate from existence, or else "existence" is being assessed according to itself, which is meaningless. This independent consistency requires that the criteria by which meaning is assigned, in some way "exist" independently from existence. This contradicts the very definition of existence. Therefore the question "What is the meaning of existence?" is a nonsensical misuse of language, and can be neither asked nor answered. "What is the meaning of existence?" is an implicit attempt to apply the concept of "meaning" without being specific about a value system by which this meaning is assessed, i.e. pushing the word "meaning" beyond the limits within which it is generally used. Equivalently, the question can be viewed as an attempt to evaluate "existence" relative to an external reference point outside of existence. Such a reference point would, if real, render the word "existence" meaningless - assuming that "existence" is defined along the lines of "all there is", and any deviation from such a definition demands that we use a different word.

The general principle that emerges is that the detached metaphysical speculation that might, in earlier times, have resulted from a question like "What is the meaning of existence?" is unlikely to be productive unless the question itself has been analyzed, so as to ensure precision in the use of the words involved and their relationship to whatever they refer to. This sort of philosophical-linguistic analysis was favoured during most of the 20th century by the predominantly Anglo-American "analytic school", the development of which was significantly contributed to by (amongst others) Ludwig Wittgenstein. For the analytic school, philosophy was a modest discipline concerned with linguistic clarification. All that could be discovered in philosophy was already within the statements and the languages involved. The languages and the statements within them just needed to be analyzed in order to clean up confusion resulting from their misuse. Analytic philosophy was incremental and applied, using logic and linguistic analysis to re-evaluate problems from other fields, exposing pseudo-problems (for instance in science) as syntactically, grammatically or contextually meaningless. This lead to an "untangling the knot of the problem" (a phrase used by Wittgenstein and frequently repeated by other philosophers), such that the problem was not solved, but rather disappeared as meaningless.

The ultimate objective of many within the analytic school was some sort of final, logically perfect language. The implicit idea was that the distinction between meaningless and meaningful language could be cleared up once and for all - in effect allowing philosophy to "stop", or at least confine itself to applied linguistic clarification rather than sweeping metaphysical speculation. The related idea of a unification of logic and mathematics (or more explicitly, the attempt to ground the "foundations" of mathematics in logic) derives in large part from theoretical/formal innovations, independently made by Russell[5] and Frege[6], in the late 19th and early 20th centuries. The common theme among these innovations was that mathematical entities were, below the surface, logical entities; that mathematics was effectively part of logic. Russell and Frege independently proposed the argument that numbers (a mathematical concept) are classes (a logical concept). The argument for this kind of homology was, roughly, that each natural number was a class/set with the appropriate number of elements. One can define an empty set with no elements in it, (which can be argued to correspond to zero), and "successor" operations by which new elements are added to existing sets. Through these successor operations the rest of the natural numbers can be recursively generated from the empty set. In other words, if you have sets, and you have the ability to include things in sets (including other sets), you can produce the natural numbers. Any specific number was regarded as an equivalence class, the class of all entities (of size n for any natural number n) that could be put into a $1:1$ correspondence with other entities within that class. Thus, any given number was an entity that could be put into a one to one correspondence with another instance of that same number. This supposed tie between the number concept, and the concept of $1:1$ correspondence between sets containing the same number of elements, presumably negated (in the minds of Russell and Frege) any criticism relating to the circularity of defining a number as a class with that "number" of elements within it.

This sort of homology between logical ideas (including things in categories, and generating new categories by operations on existing ones), and foundational mathematical ideas like "what is a number?", strengthened Russell's hope for a final, universal, logical calculus. But its glaring circularity also created confusions like Russell's paradox ("Is the class of all classes that are not members of themselves a member of itself?"). This whole line of thinking was based around the idea of providing a reliable and meaningful answer to questions of the form "What is a number?" Obviously, such thinking is founded on the supposition that answers to such questions are both possible and desirable. Wittgenstein's contribution to this school of thought can be represented by the argument that the question "What is a number?", because it refers to mathematical concepts but is at the same time mathematically useless, meaningless - and can therefore be neither asked nor answered.

The source of the confusion that the analytic school sought to clear up is, arguably, a disconnect between the rules of language and the psychological inclination to use language to ask such (borderline meaningless) questions. That is, the feeling of the need to ask something, combined with the feeling of not quite having access to the appropriate words or linguistic tools. The analytic methodology, of showing that the language does not map on to the question, by strict application of linguistic rules, can clean up the misuse of language. But such methodology is completely dismissive of anything behind the subjective inclination to ask such questions. A fundamentally different kind of philosophy, arguably more receptive to such subjective inclinations, was undertaken by the "continental school" (most of whom derived, as the name suggests, from continental europe, although it should be mentioned that German philosophers in particular made significant contributions to both schools). Continental philosophy viewed the analytic approach as having fundamentally missed the point of what philosophy should be. Heidegger was amongst the most significant contributors to this line of thinking, and built his whole philosophy around the supposition that the question "What is the meaning of being?" which is near inseparable from the (arguably) nonsensical "What is the meaning of existence?"[7]

Continental philosophers tended to produce large, holistic, all - encompassing and often "revolutionary" works of prose, based on analyses of previous, equally long-winded texts, expressing an unapologetically panoramic, and arguably unapologetically subjective, view of the human experience. They regarded this process as a fundamentally important aspect of striving for and attaining meaning, one that would not be aided by endless grammatical/syntactical analysis. In some sense therefore, one might say that the continental school regarded the very purpose of philosophy to be the posing of questions that could, strictly speaking, be considered to be meaningless - because this was the very property that made such questions philosophically important. The open endedness resulting from the linguistic ambiguity was exactly what indicated the potential for new discoveries as a result of detached reflection. By contrast, an analytic approach was, by definition, restricted to the application of pre-existing meaning criteria to language, arguably making it impossible for this approach to produce new criteria (just apply those that existed). The basic divide is perhaps best illustrated by the rumour that the analytic and continental schools were each referred to by the other as (respectively) "linguists" and "poets." Both labels were presumably intended in a less than complementary sense.

These issues boil down to the presumed relationship between language, thought, knowledge, and experience of the world. Propositions are constructed in language and may be true or false, assuming that some sort of criteria by which to assign truth and falsity can be agreed upon. In principle, given such criteria, knowledge about what is true or false may be derived from propositions. Two related fundamental distinctions exist in philosophy, discussed in greatest detail by Kant[8], between "synthetic" and "analytic" propositions[9], and between "*a priori*" and "*a posteriori*" knowledge[10]. If knowledge is "*a priori*" it is independent of experience of the external world, if knowledge is "*a posteriori*" it is derived from experience of the world. The means by which the truth or falsehood of propositions may be determined differs, depending on the content and nature of the propositions. For Kant, a synthetic proposition was one whose predicate (description) was not contained within the subject (thing being described), whereas the predicate was contained within the subject of an analytic proposition. Kant's definitive example of an analytic proposition was "all bodies are extended" (i.e. because a body is by definition composed of matter, and matter is by definition that which is extended in space). The definitive example of a synthetic proposition that Kant gave was "all bodies are heavy" (because the concept of heaviness is not inherently built into the concept of a body). Because the truth or falsehood of an analytic proposition is, by definition, already within it, all knowledge derived from analytic propositions is independent of experience - and "analytic *a posteriori* knowledge" is a contradiction in terms. Synthetic *a posteriori* knowledge is knowledge relating to a proposition that was not already contained within that proposition, and which instead depended upon experience (e.g. one would need to experience the weight of all bodies of matter in the world in order to derive knowledge from the proposition "all bodies are heavy"). Synthetic *a posteriori* knowledge includes, for example, empirically derived knowledge relevant to science. A key question for Kant, and later relevant to the analytic-continental split, was whether synthetic *a priori* knowledge was possible or not.

Synthetic *a priori* knowledge is knowledge that is independent of experience, and dependent upon propositions, but not already contained within the linguistic-semantic meaning of the constituents of those propositions. To speak extremely crudely, synthetic *a priori* knowledge is "stuff than can be just worked out", as opposed to stuff already implicitly within the linguistic tools that one might use to do the working out. All analytic (*a priori*) propositions are to some degree tautological in Kant's original conception; "all bodies are extended" because that is what the concepts "body" and "extension" mean by definition.For Kant, synthetic *a priori* knowledge was indeed a possibility. He gave examples of mathematical knowledge such as "7 + 5 = 12" or the knowledge that "The shortest distance between two points is a straight line." For Kant, the synthetic nature of knowledge concerning the truth of these claims derived from the idea that the concept of "7" does not intrinsically contain "that which can be added to 5 to produce 12", and similarly that neither the concept "point" nor "two" intrinsically contains the concept "straight line." In other words, the knowledge derived from the propositions is not intrinsically contained within the constituents of these propositions, it is derived from analysis of their combination within the proposition. What exactly is meant by one concept "intrinsically containing" another is, however, ambiguous - and this uncertainty turns out to be crucial.

The logical positivists, or logical empiricists, were a subset of analytic philosophy who redefined the synthetic/analytic distinction, viewed philosophy as broadly synonymous with the philosophy of science, and rejected the possibility of synthetic *a priori* knowledge. Positivism refers to the idea of exclusive derivation of knowledge from empirical sense data, logicism to the idea of the unification of logic and mathematics, and their use in scrutinizing sense data to produce knowledge[11]. The justification for the rejection of synthetic *a priori* knowledge rested upon a revision of what one regarded as being "already present within" language. For the logical positivists, if one has, for example, the concept of a natural number (which is necessarily present in the concept of any specific natural number), then the capacity for (say) 7 to be added to 5 to produce 12 is already present in the proposition; therefore $7 + 5 = 12$ is analytic. For the logical positivists, a true analytic proposition was essentially a tautology, the truth or falsity of an analytic proposition was built into the meaning of the terms via linguistic conventions. They regarded the truths of mathematics as indeed *a priori*, but (unlike Kant) exclusively analytic, and true by definition. Synthetic knowledge was factual, empirical, and contingent upon the external world: i.e exclusively a posteriori. For instance Carnap[12] stated that:

> "*Since empiricism had always asserted that all knowledge is based on experience, this assertion had to include knowledge in mathematics. On the other hand, we believed that with respect to this problem the rationalists had been right in rejecting the old empiricist view that the truth of '$2 + 2 = 4$' is contingent on the observation of facts, a view that would lead to the unacceptable consequence that an arithmetical statement might possibly be refuted tomorrow by new experiences. Our solution, based upon Wittgenstein's conception, consisted in asserting the thesis of empiricism only for factual truth. By contrast, the truths of logic and mathematics are not in need of confirmation by observations, because they do not state anything about the world of facts, they hold for any possible combination of facts.*"

In other words, logic and mathematics are mere languages, and any truths that they might contain are true by definition, whereas the world of "facts" is exclusively *a posteriori* and empirical. The logical positivists also dropped Kant's emphasis on the distinction between the subject and predicate in the synthetic/analytic distinction. For the logical positivists, a true analytic proposition was simply true by definition, i.e. by the meaning of its terms and/or by linguistic convention. Synthetic propositions were any propositions that were not analytic - i.e. synthetic knowledge became an umbrella term for any knowledge that did not result from mere linguistic convention. (In much the same way, "continental philosophy" was an umbrella term more widely used by successors of the logical positivists to describe any philosophy that was not analytic).

If knowledge is defined as some sort of convergence with the world, then *a priori* knowledge is impossible by definition. In this sense, critics of the logical positivists might argue that they defined out of existence any *a priori* awareness of anything. For the logical positivists, anything *a priori* was either already present in language by definition, or linguistically meaningless (as a result of the lack of any tie between language and the world). By contrast, if knowledge is defined more broadly, as some form of awareness arising in the mind, then the manipulation of language and linguistic-semantic rules (even in isolation from experience of the world), can produce knowledge that is, in some way, *a priori*. Either way, it is something of a semantic issue, dictated by one's attitude to the question as to whether knowledge can somehow come from the interaction between thought and language, if the knowledge isn't already (implicitly or explicitly) within the language. This is a subtle issue, but, importantly, it is one that is not far off the question of whether or not we can mentally/subjectively experience anything, which has sufficient coherence for us to call it "thinking", but which we could not (if we wished), express linguistically. In other words, can we think what we cannot say? In order to address this (complex and unanswered) question, it is necessary to directly compare Wittgenstein and Heidegger.

Though both born in the same year in German-speaking Europe, Wittgenstein and Heidegger superficially exhibit the differences that might be expected from their respective schools. Wittgenstein initially studied engineering and mathematics[13], and obviously had the natural aptitude to "think in pictures" possessed by many physicists and mathematicians. Wittgenstein's dense list-based style is terse to the point of being cryptic. Wittgenstein had an on-off philosophical career, matching his intermittent contempt for the subject. By contrast, Heidegger's style is ridiculously verbose and convoluted (to the point of being deliberately obscure). Heidegger initially intended to become a priest[14], and was perhaps more comfortable with subjectivity than Wittgenstein. He was a professional philosopher throughout his life, and viewed the subject as needing to start again from its foundations (as opposed to dismissing it as worthless as Wittgenstein intermittently did). There was also a striking difference in their integrity as men. Wittgenstein, despite being the heir to one of the largest fortunes in Europe[3,13] arguably seems to have been attempting to live according to an almost saint-like piety throughout his life, appearing to be "*above all, a person in search of spiritual salvation*"[15,16]. Heidegger was "*for a time a gullible and self-important Nazi*"[17]. But despite these polar differences in style, academic school, and life choices, significant convergences can be found between the implications of their philosophies.

It is the purpose of this chapter to outline these convergences (as I see them) and to argue that they have direct bearing on the concept of novelty.

It is, however, important to state initially that there is no evidence that either man thought about novelty in any great detail. Heidegger briefly mentions novelty in relation to a vulgar, shallow curiosity that characterizes a collective delusional psychology he calls the "they self." But he does not appear to have considered it metaphysically significant. Wittgenstein refers to the superstitious nature of extrapolating to the future based on the past and present, but again does not consider novelty in any depth. Nevertheless, I argue that not only is the concept of newness implicit in the work of Wittgenstein and Heidegger, but it may even provide a way around the various impasses that arise where their works diverge. I suggest that it is informative to combine the following three ideas: First, the concept of some sort of meaningful/meaningless boundary withing language (Wittgenstein). Second, the notion of pushing the boundaries of linguistic ambiguity, so as to pursue a pre-conceptual understanding that already in some way already exists within our subjectivity (Heidegger). Third, the assumption that novelty is metaphysically real. This combination gives rise to the central suggestions that I make in this chapter:

1. *Novelty provides the underlying explanation for the fact that meaningful language needs to evolve over time, and that a logical, internally consistent, meaningful/useful language, which is permanently temporally fixed and unable to evolve, is impossible.*
2. *Novelty initially enters our consciousness in a completely subjective inexpressible form, only becoming linguistically expressible later, after interaction with pre-existing thought. This, in turn, is the reason that:*
   (a) *We can sometimes "think what we cannot say"; there does not exist a consistent* 1 : 1 *relationship between thought and objective linguistic expressibility.*
   (b) *Time is a special and fundamental part of our subjectivity; it is through novelty that we most directly experience the unique properties of time.*
3. *The boundary between what we can and cannot speak of meaningfully, i.e. what Wittgenstein called "the mystical", evolves over time, due to the reality of novelty in the world.*
4. *Truth as a phenomenon is closer to disclosure/unveiling (i.e. some form of temporal concept similar to what Heidegger called aletheia) than to convergence; because novelty discloses new truths over time.*

These suggestions and the arguments surrounding them are discussed in detail at the end of this chapter. Before discussing these suggestions, I provide a conceptual summary of selected works by Wittgenstein, and of Heidegger's "*Being and Time*." (I include numerous direct quotations from the original texts in order to expand upon and justify my own summaries of key concepts. The reader who is happy to accept my reading of Wittgenstein and Heidegger at face value may wish to skip the italicized quotation sections and only read the main text of the following section). It should be emphasized that what follows is merely my reading of these philosophers, intended to provide context to my broader arguments about novelty - the reader should consult the work of a professional philosopher for a more comprehensive evaluation. Philosophers already familiar with Wittgenstein and Heidegger may wish to skip to the "Reflections and suggestions" section at the end of the chapter. Quotations are italicized, and included after the summary of the concept to which I consider them relevant. Abbreviations used are *TLP*= "*Tractatus Logico Philosophicus*", *PI*= "*Philosophical Investigations*", *OC* = "*On certainty*", *RFM*= "*Remarks on the foundations of mathematics*", "*Zettel*" *Z*, *BT*= "*Being and Time*", and *GG*= "*Groundless Grounds*", a comparison between Wittgenstein and Heidegger by contemporary philosopher Lee Braver).

Wittgenstein

The only book that Ludwig Wittgenstein published during his lifetime was *Tractatus Logico Philosophicus*[18], a piece of work that is haunting and poetic, but terse to the point of cryptic. The *Tractatus* is widely interpreted as asserting that there are no true philosophical problems, rather that "philosophical" pseudo-problems result from the application of language to a frame of reference in which that language lacks sense. Wittgenstein regarded a picture as the most useful analogy to a meaningful statement in language; a meaningful statement was in some sense a picture of the world. The linguistic frame of reference within which an accurate picture of the world could be constructed needed to be in some way "structurally similar" to the part of the world being depicted by that construction. From this idea it follows that within the frame of reference created by the rules of a language that is structurally similar to the world, certain statements are possible and others are not. If the frame of reference to which a given set of the linguistic rules refer differs from the frame of reference by which the thing being described is embedded in the world, then attempts to describe that thing in language are neither right nor wrong, but a meaningless language-world mismatch. Thus, some lines of thinking were simply blind alleys - and backing out of such blind alleys by clarifying language was all that philosophy could achieve. Along these lines, Wittgenstein's initial claim in the *Tractatus* was that he had been able "draw a limit to thought" by identifying the boundary between what could, and could not, be meaningfully said. The problem with this perspective, as will be seen, concerns the meaning of terms like "structurally similar", "frame of reference", and so forth.

The picture analogy was directly relevant to this boundary between meaningful language and nonsense: "*Though a state of affairs that would contravene the laws of physics can be represented by us spatially, one that would contravene the laws of geometry cannot.*" Spatial situations that are false when compared against real empirical physics are, despite being incorrect, still spatial situations - and can therefore be represented in a picture, in some sense meaningfully. But it is impossible to visualize a picture of something outside

3 dimensions of space. (Perhaps an advanced theoretical physicist might claim to be able to "think in" more than 3 dimensions of space, but then "imagine" would be a more appropriate word than "visualize", and they would be modifying the normal definition of the word "spatial" as it has developed in practical language). In a frame of reference that is spatial (i.e. a picture on a Cartesian (x, y, z) axis), something that is outside the laws of geometry is neither true nor false, just meaningless. Wittgenstein's basic initial argument was that meaningless philosophical pseudo-problems resulted from confusing inconsistent "spaces" and "pictures" by applying language out of context. In the *Tractatus*, pictures are facts, and pictures are only meaningful when they refer to the right kind of spaces. Words are names for things in the world, and if the words are meaningful one can point to the entities that they name on a picture, such that the relation between the objects in the state of affairs represented by the picture-fact is intrinsically similar to the relations between the real objects in the real state of affairs in the world.

In Wittgenstein's later work he had second thoughts about having completed western philosophy, in large part because he came to perceive the picture metaphor for meaningful language as incomplete. In *Philosophical Investigations*[19], Wittgenstein emphasized the interactive, evolving and social/contextual nature of what he called the "language games" - though which we attain meaningful statements via an inherently temporal, interactive process. He argued that words do not necessarily have to have fixed (or even particularly consistent) meanings, and that language games do not necessarily need to have fixed rules, but rather that the rule of language can evolve naturally over time. A rejection of the picture analogy was an important part of Wittgenstein's development of the language game concept - in the second book stating*"A picture held us captive. And we could not get outside it, for it lay in our language and language seemed to repeat it to us inexorably."* He noted how the meaning of a word or statement was inseparable from its interactive use in a social language. Wittgenstein coining the term "form of life" for the language game between one subject with an intention, a statement within a language with certain rules, and an interpretation of the statement by a second subject, again working through the (social, evolving) rules. Language games are fluid, words can mean different things to different subjects in different situations. Pinning down the "ultimate" or "objective" meaning of a word is impossible in isolation from a connection between two communicating subjects, for instance *"Here the term 'language-game' is meant to bring into prominence the fact that the speaking of language is part of an activity, or a form of life."* Statements in language went from being conceived of as pictures of facts to being tools to be used in life. This dramatic change in Wittgenstein's perspective is important. What is it about meaningful language that makes a tool a preferable analogy to a picture? I argue that one of the key differences between a tool (or a form of life, or anything else interactive) and a picture, is that a tool is used over a time interval in order to do something - whereas a picture is a fixed slice of the world whilst *time is frozen*. The central nature of this distinction will become apparent through a detailed summary of Wittgenstein's major works.

Tractatus Logico Philosophicus: Kicking away a ladder of nonsense

Wittgenstein uses the phrase "the world" where others might use the word "existence": The world is everything that is the case. He emphasizes relationships between things, not just things: The world is composed of facts (the existence of states of affairs). He develops a strong pictorial analogy for meaningfulness: A thought is a picture of a fact. He regards language (and its meaning or lack thereof) as derived from the manner in which it reflects thought: thoughts can be expressed with sense in propositions. He builds language into a hierarchical ladder that is potentially finite: Propositions are ultimately built on elementary propositions, which are simple names of things. A proposition is a truth function, a description that can be true or false (he maintains a basic binary notion of truth and falsity), of elementary propositions, and an elementary proposition is a truth function of itself. The general recipe for a meaningful description of the world is the collection of all propositions, the collection of all relationships between propositions that have some intrinsic similarity (propositional variables), and the possibility of negation of all such propositions. The totality of this general recipe for meaningful language is the general propositional form. Anything outside this category is in the realm of statements that lack meaningful constraints upon them. This is the category of which we must remain silent, because we are unable to speak of it meaningfully.

The world. An objective totality, defined by relationships rather than by discrete entities.

> *"The world is all that is the case." TLP1.*
> *"The world is the totality of facts, not of things." TLP1.1*

Pictures. The representation within language of the intrinsic structures of states of affairs in the world, i.e. a depiction of the existence of a particular state of affairs. The internal structure (the pictorial form) of the picture mirrors the internal structure of the fact in the world (the state of affairs). A thought is a picture of the world in a mind, which can be true or false, but must be of a similar form to the depicted part of the world. This internal homology is the basis of all language-world, picture-fact homology: similarity between pictorial form and the logical form of reality.

> *"A picture is a fact." TLP2.141*
> *"The fact that elements of a picture are related to one another in a determinate way represents that things are related to one another in the same*

> *way. Let us call this connection of its elements the structure of the picture, and let us call the possibility of this structure the pictorial form of the picture."* TLP2.15
> *"What any picture, of whatever form, must have in common with reality, in order to be able to depict it - correctly or incorrectly - in any way at all, is logical form, i.e. the form of reality."* TLP2.18
> *"A picture depicts reality by representing a possibility of existence and non-existence of states of affairs."* TLP2.201
> *"What a picture represents it represents independently of its truth or falsity, by means of its pictorial form."* TLP2.22
> *"The agreement or disagreement of its sense with reality constitutes its truth or falsity."* TLP2.222

States of affairs. A configuration of objects in the world, a situation. (The relationship between the objects within a state of affairs is the logical form of reality, and is what must be mirrored within pictorial form in order for picture-facts to be meaningful). A state of affairs can exist or not exist (be the case or not), this combination of existence and non-existence constitutes reality.

> *"What is the case-a fact- is the existence of states of affairs."* TLP2
> *"A state of affairs (a state of things) is a combination of objects (things)."* TLP2.01
> *" 'A state of affairs is thinkable': what this means is that we can picture it to ourselves."* TLP3.001

Objects/Things. The simplest entities, the basic, indivisible elements of reality. Objects are constituents of states of affairs within the world, and the interactive combination of objects in states of affairs, i.e. facts, are mirrored by picture-facts in language. It does not make sense to think about the thing without imagining it in the context of a state of affairs - an object literally is its potential combination with other objects in all possible states of affairs within which it may occur. Wittgenstein invokes a kind of pre-determinism for states of affairs and objects: any object has the possibility of the states of affairs within which it may occur "written into it" from the start. If we can describe the object we can describe all possible states of affairs in which it may occur, without any room for subsequent undiscovered states of affairs emerging. (It is notable that this distinctly anti-novelty view is inconsistent with Wittgenstein's later philosophy).

> *"Each thing is, as it were, in a space of possible states of affairs. This space I can imagine empty, but I cannot imagine the thing without the space."* TLP2.013
> *"In logic nothing is accidental: if a thing can occur in a state of affairs, the possibility of the state of affairs must be written into the thing itself."* TLP2.012
> *"If I know an object I also know all its possible combinations in states of affairs. (Every one of these possibilities must be part of the nature of the object). A new possibility cannot be discovered later."* TLP2.0123

Substance. Form and content; what there would be without specific states of affairs, i.e. a kind of prerequisite reference point underlying the existence of all specific states of affairs, and being necessary for their existence. Substance is the grounding for the possibility of existence of specific facts. (Wittgenstein does not say what if any grounding substance itself is supposed to have).

> *"Substance is what subsists independently of what is the case."* *TLP*2.024
> *"It is form and content."* *TLP*2.025

Thoughts. Mental pictures of states of affairs, i.e. the manifestation of picture-fact representations within mind. Wittgenstein (at least in the *Tractatus*) appears to view logic and (expressible) thought as almost synonymous, and logical possibility as a prerequisite for thought. That is, if we can think something, that thing is "logically possible" and therefore (one assumes) logically representable. Thought, logic and reality are overlapping homologous categories with the same fundamental structure. There is no such thing as an illogical thought.

> *"A logical picture of facts is a thought."* *TLP*3
> *"The totality of true thoughts is a picture of the world."* *TLP*3.01
> *"A thought contains the possibility of the situation of which it is the thought. What is thinkable is possible too."* *TLP*3.02
> *"Thought can never be of anything illogical since, if it were, we should have to think illogically."* *TLP*3.03
> *"It is as impossible to represent in language anything that contradicts logic, as it is in geometry to represent by its co-ordinates a figure that contradicts the laws of space, or to give the co-ordinates of a point that does not exist."* *TLP*3.032

Propositions. Expressions of thoughts. Descriptions within language, of thoughts within mind, about states of affairs within the world. A proposition is something that can be true or false; propositions are related to states of affairs by whether or not a they truthfully describe the state of affairs that they point to. The intrinsic structure of a proposition is necessarily similar (in "form") to whatever the proposition refers to. The sense, i.e. the meaning, of a proposition remains only a possibility until the proposition is referred to the world.

> *"The substance of the world can only determine a form, and not any material properties. For it is only by means of propositions that material properties are represented - only by the configuration of objects that they are produced."* *TLP*2.0231
> *"In a proposition a thought finds an expression that can be perceived by the senses."* *TLP*3.1

Propositional signs/expressions/symbols. A propositional sign is an interpretable expression of a proposition, and thus of a thought. I.e. when a person states a proposition, they project their own thoughts into somebody else's perception. A propositional sign is therefore a linguistic device through which thoughts are expressed within propositions. A propositional symbol is a meaningful expression of a proposition. (Crudely, a propositional sign is part of the language of propositions, within which facts can be stated, whereas a propositional symbol is a language-world relationship with meaning, through which facts can be ascertained to be true or false. For a more sophisticated discussion of the sign /symbol distinction, see McManus, below).

> *"I call the sign through which we express a thought a propositional sign. - And a proposition is a propositional sign in its projective relation to the world."* TLP3.12
> *"What constitutes a propositional sign is that in it its elements (the words) stand in a determinate relation to one another. A propositional sign is a fact."* TLP3.14
> *"In a proposition a thought can be expressed in such a way that elements of the propositional sign correspond to the objects of the thought."* TLP3.2
> *"I call any part of a proposition that characterizes its sense an expression (or a symbol). (A proposition is itself an expression). Everything essential to their sense that propositions have in common with one another is an expression. An expression is the mark of form and content."* TLP3.31

Sense. Meaning, the truth or falsehood of thought-language-world relations. The sense of a proposition is the meaning of that proposition. A proposition cannot be true or false in and of itself, only though its relation to the configuration of objects within states of affairs in the world.

> *"What a picture represents is its sense."* TLP2.221
> *"A proposition includes all that the projection includes, but not what is projected. Therefore, though what is projected is not itself included, its possibility is. A proposition therefore, does not actually contain its sense, but does contain the possibility of expressing it. ('The content of a proposition' means the content of a proposition that has sense). A proposition contains the form, but not the content, of its sense"* TLP3.13
> *"A proposition shows its sense. A proposition shows how things stand if it is true. And it says that they do stand."* TLP4.022

Names. Simple labels in language for objects in the world. A name is a word used to denote an object. Names are constituents of propositions but are incapable of describing states of affairs in themselves, because they are more simple than the propositions of which they form part, and cannot convey relationships without the wider structure of these propositions. Names can refer to objects but not states of affairs, because the latter are more complex.

> *"Only facts can express a sense, a set of names cannot."* TLP3.142
> *"Situations can be described but not given names. (Names are like points; propositions like arrows-they have sense.)"* TLP3.144
> *"A name means an object. The object is its meaning 'A' is the same sign as 'A'."* TLP3.203

Elementary propositions. The simplest possible propositions, which assert only the existence of states of affairs, and consisting only of a connection between names.

> *"The simplest kind of proposition, an elementary proposition, asserts the existence of a state of affairs."* TLP4.21

> *"It is a sign of a proposition's being elementary that there is no elementary proposition contradicting it."* TLP4.211
> *"An elementary proposition consists of names. It is a nexus, a concatenation, of names."* TLP4.22
> *"It immediately strikes one as palpable that the introduction of elementary propositions provides the basis for understanding all other kinds of proposition. Indeed the understanding of general propositions palpably depends on the understanding of elementary propositions."* TLP4.411

Propositional variables. The generic form of a family of propositions in which a particular expression may occur. An expression, an interpretable projection of a proposition that conveys sense about a state of affairs, can be related to a collection of particular propositions, within which this expression may occur. This collection of propositions can be expressed in generalized form as a variable, the particular values of which are the particular propositions containing the expression. Propositional variables refer to what a given family of propositions have in common, i.e. act as a signifier for the formal concept underlying that family of propositions, which shares the capacity for inclusion of a particular expression.

> *"An expression presupposes the forms of all the propositions in which it can occur. It is the common characteristic mark of a class of propositions."* TLP3.311
> *"It is therefore presented by means of the general form of the propositions that is characterizes. In fact, in this form, the expression will be constant and everything else variable."* TLP3.312
> *"Thus an expression is presented by means of a variable whose values are the propositions that contain the expression. (In the limiting case the variable becomes a constant, the expression becomes a proposition.)"* TLP3.313
> *"If we turn a constituent of a proposition into a variable, there is a class of propositions all of which are values of the resulting variable proposition. In general, this class too will be dependent on the meaning that our arbitrary conventions have given to parts of the original proposition. But if all the signs in it that have arbitrarily determined meanings are turned into variables, we shall still get a class of this kind. This one, however, is not dependent on any convention, but solely on the nature of the proposition. It corresponds to a logical form- a logical prototype."* TLP3.315

Formal concepts. A implicit, intrinsic concept that is "pointed at" by a propositional variable, but not expressed by any of the associated propositions. A formal concept is something underlying a particular collection of linguistic entities, which "shows itself" in expressions involving them, but cannot be explicitly described in a meaningful way by such expressions. A variable, the property that a class of propositions shares so as to be able to include a particular expression, is a sign, a linguistic pointer to, a formal concept. Formal concepts are the prerequisites for specific instances of formal language, but cannot be directly described within these instances of language. For example, the formal concept "object" is reflected when we use it in a derivative, practical sense, by saying (for instance) "There are two stones on the ground", in that this could be a meaningful proposition, which related (truthfully or not) to the world. But if we tried to construct a meaningful proposition about "objects" in general, e.g. "objectness is manifest in reality" we would end up with meaningless language. This is because the formal concept "object" underlies the specific propositions in which we talk about particular objects in particular situations. Meaningfulness is restricted to this specificity - we move out of meaningful language if we try to speak about formal concepts in themselves. For instance, within the frame of reference of such propositions we cannot talk meaningfully about objects without being specific. Nevertheless, the formal concept "object" is is shown by such propositions, even though it can't be meaningfully described by them. A formal concept is thus something that we intuitively recognize as a similarity amongst a collection of expressions, that is a grounding prerequisite for the existence of such expressions, but is wider in scope than the frame of reference in which they have meaning. It is perhaps with the introduction of this idea that Wittgenstein begins to move beyond what the impatient reader might construe as a long-winded and pedantic labeling exercise, and starts to "say" something about the unsayable that is potentially profound.

> *"We now talk about formal concepts, in the same sense that we speak of formal properties. (I introduce this expression in order to exhibit the source of the confusion between formal concepts and concepts proper, which pervades the whole of traditional logic.) When something falls under a formal concept as one of its objects, this cannot be expressed by means of a proposition. Instead it is shown in the very sign for this object. (A name shows that it signifies an object, a sign for a number that it signifies a number, etc.) Formal concepts cannot, in fact be represented by means of a function, as concepts proper can, For their characteristics, formal properties, are not expressed by means of functions. The expression for a formal property is a feature of certain symbols. So the sign for the characteristics of a formal concept is a distinctive feature of all symbols whose meanings fall under the concept. So the expression for a formal concept is a propositional variable in which this distinctive feature alone is constant."* *TLP*4.126
>
> *"Every variable is the sign for a formal concept. For every variable represents a constant form that all its values possesses, and this can be regarded as a formal property of those values."* *TLP*4.1271
>
> *"Thus the variable 'x' is the proper sign for the pseudo concept object. Wherever the word 'object' ('thing' etc) is correctly used, it is expressed in conceptual notation by a variable name. For example, in the proposition 'There are 2 objects which...', it is expressed by '$(\exists x, y)$...'. Wherever it is used in a different way, that is as a proper concept-word, nonsensical pseudo propositions are the result. So one cannot say, for example, 'There are objects', as one might say 'There are books'. And it is just as impossible to say, 'There are 100 objects' or 'There are $\aleph_0$objects'. And it is nonsensical to speak of the total number of objects. The same applies to the words 'complex', 'fact', 'function', 'number', etc. They all signify formal concepts, and are represented in conceptual notation by variables, not by functions or classes (as Frege and Russell believed). '1 is a number', 'There is only one zero' and all similar expressions are nonsensical, (It is just as nonsensical to say 'There is only one 1' as it would be to say, '2 + 2 at 3o'clock equals 4')."* *TLP*4.1272
>
> *"A formal concept is given immediately any object falling under it is given. It is not possible, therefore, to introduce as primitive ideas objects belonging to a formal concept and a formal concept itself. So it is impossible, for*

example, to introduce as primitive ideas both the concept of a function and specific functions, as Russell does; or the concept of a number and particular numbers." TLP4.12721
"To ask whether a formal concept exists is nonsensical. For no proposition can be the answer to such a question. (So, for example, the question 'Are there unanalyzable subject-predicate propositions?' cannot be asked.)" TLP4.1274

Form. The property of a series that causes it to be ordered by an internal relation shared by its elements. In the example below, a stands in relation to b, there exists x such that a stands in relation to x and x stands in relation to b, there exists x and y such that a stands in relation to x and x stands in relation to y, y stands in relation to b. In other words, the propositions showing how all these terms can be positioned relative to each other via the relation R implies that they share a common ordering property that Wittgenstein calls form.

"I call a series that is ordered by an internal relation a series of forms. The order of the number series is not governed by an external relation but by an internal relation. The same is true of the series of propositions $'aRb, \exists x : aRx.xRb, \exists x, y : aRx.xRy.yRb$*, and so forth."* TLP4.1252

Possibilities of existence. Combinations of existence/non-existence of different particular states of affairs. I.e. for each state of affairs in a set, each one can exist or not, the combination of all the states of affairs that do exist (constitute true facts about the world), is one possibility of existence. (($\binom{n}{v}$) means "number of combinations of v objects chosen from a collection of n objects", ($\binom{n}{v}$) $= \frac{n!}{(n-v)!v!}$. $\sum$ denotes a sum of anywhere between the lower and upper subscripts, so $\sum_{v=0}^{n}(\binom{n}{v})$ means the sum total number of combinations of objects within collection of all possible sizes from a total pool of n objects).

"For n states of affairs, there are $K_n = \sum_{v=0}^{n}(\binom{n}{v})$ *possibilities of existence and non-existence. Of these states of affairs any combination can exist and the remainder not exist."* TLP4.27
"There correspond to these combinations the same number of possibilities of truth -and falsity- for n elementary propositions." TLP4.28

Truth possibilities. The existence (or not) in the world of particular states of affairs referred to by propositions. (I.e. possibilities of truth or falsehood when propositions are compared to the world).

"Truth possibilities of elementary propositions mean possibilities of existence and non-existence of states of affairs." TLP4.3
"Truth possibilities of elementary propositions are the conditions of the truth and falsity of propositions." TLP4.41

Truth conditions. The expression of agreement (or not) between the truth possibilities of the constituent elementary propositions that occur within a complex, non-elementary proposition. I.e. Wittgenstein asserts that all propositions are complexes of elementary propositions expressed in some form of relationship to each other, and that the meaning of a proposition is the result of some sort of comparison between the truth possibilities of the elementary propositions of which it is a complex. All non-elementary propositions are, when compared with each other, on a continuum between a tautology (all its elementary propositions agree in terms of their truth possibilities) and a contradiction (the truth possibilities of their elementary propositions make it impossible for them to ever agree). (This implies that if we really know all elementary propositions, we can construct an exhaustive, true picture of the world by combining them in more complex propositions).

> "*The expression of agreement and disagreement with the truth-possibilities of elementary propositions expresses the truth conditions of a proposition. A proposition is the expression of its truth conditions....*" *TLP*4.431
> "*Among the possible groups of truth-conditions there are two extreme cases. In one of these cases the proposition is true for all the truth possibilities. We say that the truth conditions are tautological. in the second case the proposition is false for all the truth possibilities: the truth-conditions are contradictory. In the first case we say the proposition is a tautology; in the second, a contradiction.*" *TLP*4.46
> "*Propositions comprise all that follows from the totality of all elementary propositions (and, of course, from its being the totality of them all). (Thus, in a certain sense, it could be said that all propositions were generalizations of elementary propositions).*" *TLP*4.52
> "*A proposition is a truth-function of elementary propositions. (An elementary proposition is a truth-function of itself).*" *TLP*5

Truth grounds. The truth possibilities of the elementary propositions within larger propositions that (if true) are sufficient to make the larger proposition true. I.e. the subset of the truth possibilities of a particular proposition that is sufficient to render that proposition true.

> "*Elementary propositions are the truth arguments of propositions.*" *TLP*5.01
> "*....I will give the name truth grounds of a proposition to those truth possibilities of its truth-arguments that make it true.*" *TLP*5.101
> "*If all the truth-grounds that are common to a number of propositions are at the same time truth grounds of a certain proposition, then we say that the truth of that proposition follows from the truth of the others.*" *TLP*5.11

Propositional structure. Crucially, Wittgenstein asserts that the similarity in Truth-grounds leads to a structural similarity between propositions that agree with each other. I.e. if propositions agree, they just "show" that they agree intrinsically. Truth appears, for Wittgenstein, very close to structural correspondence between propositions that refer to the same states of affairs. The meaning of propositions is within them, in how they are constructed, as opposed to being contained in any comparative laws by which propositions can be arranged relative to each other. Elementary propositions, referring only to the names of objects in the world, cannot be deduced from each other, this is why they are elementary. Wittgenstein makes a comment about deducing the situations of the future from those of the present, which is connected to our inability to make inferences from one elementary proposition to another, and also to the validity of applying rules of induction over time. The latter point could be read as leaving open the possibility of genuine novelty.

> *"When the truth of one proposition follows from the truth of another, we can see this from the structure of the propositions."* TLP5.13
> *"If the truth of one proposition follows from the truth of another, this finds expression in the relations in which the forms of the propositions stand to one another: nor is it necessary for us to set up these relations between them, by combining them with one another in a single proposition; on the contrary, the relations are internal, and the existence is an immediate result of the existence of the propositions."* TLP5.131
> *"If p follows from q, I can make an inference from q to p, deduce p from q. The nature of the inference can be gathered only from the two propositions. They themselves are the only possible justification of the inference. 'Laws of inference', which are supposed to justify inferences, as in the works of Frege and Russell, have no sense, and would be superfluous."* TLP5.132
> *"One elementary proposition cannot be deduced from another."* TLP5.134
> *"There is no possible way of making an inference from the existence of one situation to the existence of another, entirely different situation. There is no causal nexus to justify such an inference."* TLP5.136
> *"We cannot infer the events of the future from those of the present. Superstition is nothing but belief in the causal nexus."* TLP5.1361
> *"The structure of propositions stand in internal relations to one another."* TLP5.2

Truth operations. An operation is what is done to the constituents of a proposition in order to produce that proposition. A truth operation is an operation that acts upon elementary propositions to produce a proposition that relates to states of affairs, and can therefore be true or false.

> *"...we can represent a proposition as the result of an operation that produces it out of other propositions (which are bases of the operation)."* TLP5.21
> *"An operation is the expression of a relation between the structures of its result and of its bases."* TLP5.22
> *"Truth functions of elementary propositions are results of operations with elementary propositions as bases. (These operations I call truth-operations)."* TLP5.234

General propositional form. The generic structure of all possible meaningful descriptions of the world. The totality of all propositions $\bar{p}$, all propositional variables $\bar{\xi}$, and the negation of all values of all propositional variables $N(\bar{\xi})$, all expressed in their most parsimonious and therefore most meaningful form. The form of all propositional variables, truth and negation of all of their truth possibilities, and intrinsic relationships between the truth grounds of all propositions is a basic form that covers any and all possible meaningful description of the world in language. The general propositional form is a structure within the logic of our language, which how things must be in order for us to talk about the world in a meaningful way, i.e. what an internally consistent system for discussing the truth must be like. This is the structural grounding of logical language and it reflects, but does not and cannot directly describe, the structure of reality.

> "*Occam's maxim is, of course, not an arbitrary rule, nor one that that is justified by its success in practice: its point is that unnecessary units in a sign language mean nothing. Signs that serve one purpose are logically equivalent, and signs that serve none are logically meaningless.*" TLP5.47321
> "*....$N(\bar{\xi})$ is the negation of all values of the propositional variable $\bar{\xi}$.*" TLP5.502
> "*If $\bar{\xi}$ has only one value then $N(\bar{\xi}) = \sim p$ (not p): if it has two values, $N(\bar{\xi}) = \sim p. \sim q$ (neither p nor q).*" TLP5.51
> "*The general form of a truth-function is $[\bar{p}, \bar{\xi}, N(\bar{\xi})]$. This is the general form of a proposition.*" TLP6
> "*What this says is just that every proposition is a result of successive applications to elementary propositions of the operation $N(\bar{\xi})$.*" TLP6.001
> "*The propositions of logic describe the scaffolding of the world, or rather they represent it. They have no 'subject matter'. They presuppose that names have meaning and elementary propositions sense; and that is their connection with the world. It is clear that something about the world must be indicated by the fact that certain combinations of symbols-whose essence involves the possession of a determinate character-are tautologies. This contains the decisive point. We have said that some things are arbitrary in the symbols that we use and some things are not. In logic it is only the latter that we express: but that means that logic is not a field in which we express what we wish with the help of signs, but rather one in which the nature of the absolutely necessary signs speaks for itself. If we know the logical syntax of any sign-language, then we have already been given all the propositions of logic.*" TLP6.1254

The mystical. An inexpressible, intuitive, subjective feeling provoked by existence. Arguably related to the apparent finite nature of existence combined with a feeling that existence is necessarily related to something "off the scale" of what can be described by languages that are necessarily constrained by their structural similarity to what does exist. The mystical is the structure of the world that is reflected by the structure of meaningful language, but which cannot be described by that language, precisely because it grounds linguistic structure; that which cannot be put into words but only shown, and must therefore be passed over in silence.

> "*The facts all contribute only to setting the problem, not to its solution.*" TLP6.4321
> "*It is not how things are in the world that is mystical, but that it exists.*" TLP6.44
> "*We feel that even when all possible scientific questions have been answered, the problems of life remain untouched. Of course there are then no questions left, and this itself is the answer.*" TLP6.52
> "*There are, indeed, things that cannot be put into words. They make themselves manifest. They are what is mystical.*" TLP6.522
> "*My propositions serve as elucidations in the following way: anyone who understands me eventually recognizes them as nonsensical, when he has used them-as steps- to climb up beyond them. (He must, so to speak, throw away the ladder after he has climbed up it). He must transcend these propositions, and then he will see the world aright.*" TLP6.54

"*What we cannot speak about we must pass over in silence.*" *TLP7*

So for Wittgenstein, anything we can say can be said clearly, but language is built on names, which are simple labels by which we refer to objects, within the world, of which we are inextricably a part, and natural human communication can be summed up via something along the lines of:

$$\begin{array}{ccccc} state & truth/falsity & & expression/symbol & \\ of & \longleftrightarrow & proposition & \longrightarrow & perception \\ affairs & thought & & language & \end{array}$$

In asking something like "What is an object?" we step outside the realm in which the language refers to anything - because the concept "object" is unavoidably defined as part of the means by which language names things in the world. To describe in language how this naming process works, we would need to use language to step outside the world-structure that grounds language, which is obviously impossible. All that philosophy, or any attempt at the construction of meaningful language, can do, is reflect the world from inside it, rather than attempting to "objectively" describe it from the outside. The meaningless unaskable problems therefore include things like "what is an object?", "what is a name?" etc. These unaskable meaningless questions do, however, pertain to the foundational elementary propositions that Wittgenstein builds his whole thesis on. In this sense the *Tractatus* undermines its own foundations - one cannot ask why states of affairs are combinations of objects, or why we refer to objects with names, etc. Language reflects the structure of reality, and all we can do is silently notice. This is Wittgenstein's fundamental claim, but it is one that ultimately he just tautologically asserts, rather than proves or logically justifies.

In his introduction to the *Tractatus*, Bertrand Russell commented that, for a man who dismisses so much philosophy as mere meaningless language, Wittgenstein managed to use a good deal of meaningless language himself: "***What causes hesitation is the fact that, after all, Mr Wittgenstein manages to say a good deal about what cannot be said, thus suggesting to the skeptical reader that possibly there may be some loophole through a hierarchy of languages , or some other exit.***" Wittgenstein would presumably have responded to this by suggesting that Russell had missed the point, and that a scaffold of meaningless language was acceptable once kicked away, that what matter is the subsequent view of the structure of reality. If your claim gets you to the truth, the materials of which the ladder is composed are unimportant. By trying to build the hierarchy of languages that Russell suggests, we cannot avoid a regress through "What aspects of the world can be named as objects?", "Why is the world describable?" etc, all the way to "Why is there anything?" - all of which Wittgenstein would dismiss as unaskable because they are outside the logic of our language. This is Wittgenstein's point when he says "*It is not how things are in the world that is mystical, but that it exists.*" (which, incidentally, is Heidegger's starting point). What we are trying to get at with such questions is indeed important, but the questions simply cannot be answered. Perhaps this, for Wittgenstein, is the important thing, his initial intuition being that the inexpressible (which he perceived as synonymous with the mystical) was in fact more important than what could be meaningfully expressed: "*my work consists of two parts: of the one which is here, and of everything which I have not written. And precisely this second part is the important one.*"[20]. Similarly, biologist C.H. Waddington, an acquaintance of Wittgenstein, later commented:

> *"No one who met him could easily doubt that Wittgenstein was a tragic figure of some magnitude...[they] failed to realize that what he had to convey was essentially a poetic awareness of the otherness of reality; he was led, presumably by historical accidents of which I am ignorant, to enshrine his message in a logico-philosophical jargon.....[which] resulted in him being taken for a philosopher - an identification by which he was at first flattered, later filled by apprehension and a sense of guilt...*"[21]

Assume we listen to Wittgenstein's advice about "kicking the ladder away" because it shows itself to be nonsense. Assume we do not get too concerned about what exactly he means by an object, by a state of affairs, by a name, etc. In other words, assume we do not ask for an ending, foundational point to the indefinite recursion by which one such term is defined relative to another. Then we are left with what shows itself in meaningful language, what is left when we stop worrying about words, and simply engage with the world. We move towards what Wittgenstein seems to be getting at by "formal concepts", fundamental constraints on existence that just show themselves through linguistic structure. We arrive at "General propositional form", the underlying fabric of how the world necessarily must be in order to be describable at all. If we even begin to ask questions like "Why is general propositional form there, why does it exist?", i.e. "Why does language necessarily reflect an underlying structure of relationships within the world in order to be meaningful?" or "Why do we need to implicitly assume that names have meaning, and the elementary propositions have sense in order for logical language to reflect the world?" then, straight away, we get stuck- because we are stepping outside the logic of our language, and must therefore pass the question over in silence. All that philosophy can do is remind us of this, and highlight the natural logical structure that our language already has.

This is Wittgenstein's a daringly simple advance, and it is all well and good. But how exactly do we tell the difference between meaningful and meaningless language? Who is to decide which statements have meaning and which do not, and by which criteria? How do we know when we have reached the point at which we must pass the question over in silence, simply appreciating the unbounded structure of the world that makes itself manifest? Wittgenstein's definition of a meaningful statement would presumably be along the lines of "a linguistic proposition, the structure of which mirrors the structure of the aspect of the world it is describing." If we asked how we are supposed to know when the world's deep structure is being correctly mirrored and when it is not, perhaps we might get a response along the lines of 4.022 "*A proposition shows its sense. A proposition shows how things stand if it is true. And it says that they do stand..*" We might then ask how we are supposed to know when a true proposition is authentically showing itself, the answer to which is not immediately obvious. Eventually we would probably get an appeal to pure intuition; we would get a "sense" when a formal concept, or the wider general propositional form of logic, was making itself manifest in our language. Thus we see the journey Wittgenstein deliberately attempts to make from logician/philosopher to mystic/prophet of silence. Although few men of recent times seem more genuinely worthy of the role of philosopher-mystic than Wittgenstein, he appears to have (initially) been oblivious to the fact that human language evolved for reasons far more mundane and practical than arriving at a foundational appreciation of the structure of reality. The problem with kicking away the ladder of language is that we are stuck at the top appreciating the mystical, and we cannot go up and down to navigate the nuts and bolts of the mapping between language and the world. We need a consistent, practical way of deciding which linguistic statements are meaningful and at which are not. Whatever subjective appeal there may or may not be to passing the mystical over in silence, we need to hold on to the ladder of language, because we need a practical way in which decide what is, and is not, meaningful language. Wittgenstein seems to have thought along these lines in his later work.

Philosophical investigations: Meaning is use, use is contextual and evolving

In the posthumously published *Philosophical Investigations*, Wittgenstein changed his mind about the *Tractatus* having been the completion of western philosophy. He rejected the "picture theory" of meaning, and developed a more contextual and nuanced view of meaningful language. This involved abandoning any notion of an absolutely consistent, unassailably reliable connection between a particular proposition in language, and either the part of the world being described or the mind doing the describing. Wittgenstein instead emphasized the contextual and interactive nature of language and the implicit, social nature of its rules. In this conception, the meaning of a word is nothing more than its use within a communication exercise with another person, a "language game." The understanding of a word, a sound, or a symbol, is not so much a mental/conscious process, as an ability, akin to the ability to successfully play a game according to that game's rules. Just as is the case with the rules of a game, there is no specific need to invoke a relation between the rules by which the players conduct themselves and either the ideas in their minds or the physical aspects of the world with which they interact. A particular game "just is" played according to its rules. Were it not so, it would not be that specific game that was being played. As Grayling puts it "*...there is no answer to the question 'Why [in chess] does the king move only one square at a time?', for if it is chess one is playing then that is simply the rule.*"[3] An "Objective meaning" of words that can be defined in a detached, esoteric way, separate from the language game, is not something that is particularly relevant to that language game, or even necessarily real. Rules in language games are evolving conventions, rather than "objective" coercive mechanisms that "make" the rule follower engage with the game[3].

Thus, the rules of language games are manifest (as opposed to abstract/theoretical) social conventions, which require no external referent, no grounding. What is "true" or "false" is decided upon by what human beings say is true and false, with reference to rules that they must agree upon and refer to in order to define this truth or falsity. The whole interactivity, between people, their language games and its rules, and the meaning derived from participation in these games with reference to these rules, is what Wittgenstein calls a "form of life." Truth and falsehood are agreement (or lack thereof) between particular forms of life - and that is all, there is no need to refer to anything outside the language game, or to look for a "deeper" definition. In stark contrast to the confident hierarchically structured set of deliberately enigmatic assertions in the *Tractatus*, *Philosophical Investigations* is filled with examples of situations in which the meaning of words is plagued with ambiguity. Wittgenstein presents various thought experiments and attempts to point the reader in the direct of his conclusion (rather than aggressively declaring it to be true as he does in the *Tractatus*). The general principle that emerges is that any instance of meaningful language, no matter how self evident or internally consistent it may appear, becomes ambiguous, and in most cases outright meaningless, in at least some situations. The development of meaningful language is a practical matter, one that we just get on with - not something that we derive logically, or by criteria that are prior to the actual use of language.

For instance, suppose a person is teaching a young child who is learning language for the first time the meaning of (say) the word "cat", and their pet cat happens to be black. They point at the physical object, make the appropriate sound, and hope that after sufficient repetition the child will understand the association between the word "cat" and the animal. If "names" are words used to point to "objects", the name "cat" denotes the physical object, the animal. Once we have enough names we can build more complex propositions, and eventually construct a logically perfect language, the structure of which will mirror that of the world (and perhaps tell us something compelling). But suppose the person is teaching the child the meaning of the word "black" (and the cat happens to be the only black object in the room). Other than the word used, the process of repeating the word, pointing at the object, and hoping that after enough repetition the child will understand, would be no different. Thus, in this real, practical instance of the use of language, do both words name the same object? Clearly, the answer is "Yes and no", and it doesn't particularly matter either way. It is an obviously irrelevant and unanswerable question to ask whether the named object is "black-ness", of which "cat-ness" is a derivative property, or the other way around. And, crucially, the reason that this would be a silly thing to ask is that this question is of no relevance to teaching a child what a "black cat" is. The *Tractatus*' propositional hierarchy of increasing complexity, which shows the underlying structure of the world as it makes itself manifest, simply is not an accurate representation of how human beings learn and use language. There is merely an interactive communication exercise between two conscious minds. In Wittgenstein's later work he had reached the view that meaning was nothing more than use in such a communication exercise, a form of life. His themes can be categorized according to the following implicit arguments, most of which are consistent with the basic spirit of the *Tractatus*, but which attempt to correct the book's finality:

Philosophy as "language therapy." Meaningless pseudo-problems arise from the misuse of language, and the role of philosophy is to "treat" such meaninglessness by linguistic clarification. Nonsense results from language being used for a purpose that is too far removed from that for which it was initially developed. Philosophy is a therapeutic exercise, in which the philosophical impulse is "cured" by linguistic clarification. Such clarification causes the "knot of the problem" to become "untangled." Ordinary language is developed for practical purposes, and when such language is "refined" beyond the context in which those practical purposes are relevant (as happens in the development of logical-philosophical "idealized" language) it ceases to be useful. In the case of logic, one might say that it simply shows the structure that language necessarily had to have in order to develop usefully in the first place, as opposed to uncovering anything new. Such "refinement" is the result of excessive scrutinization of words, in isolation from their use, something that results in "philosophical" pseudo-problems.

> "*The results of philosophy are the uncovering of one or another piece of plain nonsense and of bumps that the understanding has got by running its head up against the limits of language.*" *PI*119
>
> "*The fundamental fact here is that we lay down rules.....and that when we follow the rules, things do not turn out as we had assumed....we are therefore as it were entangled in our own rules.....*" *PI*125
>
> "*The philosopher's treatment of a question is like the treatment of an illness.*" *PI*255

Meaning is Use. Definitions of words are (at best) implicit in use, the meaning of words cannot be defined in an abstract way outside the context in which those words are used. There is no way of defining an abstract linguistic form devoid of any connection with the practicalities of the context in which communication happens. A word means something if the linguistic interaction is successful -with "success" being contingent on the intentions of the individuals interacting through language: A word means something if one can use it, where "use" has a practical and specific, not esoteric or theoretical, definition. The only way that language can, strictly speaking, describe the world at its most basic level is to name things. (There is an obvious parallel with the *Tractatus* here). But to ask what a name is is the same as asking what a "simple" denoted by names is. The question is meaningless; there is no definition of names, or the "primary simples" within the world that names denote. All that can be said is that a particular substance is "pointed at" using a particular name. At the most basic level the meaning of any aspect of language is nothing more than its practical utility. Importantly, this concept of use is inherently temporal.

> "*......the meaning of a word is its use in the language. And the meaning of a name is sometimes explained by pointing to its bearer.*" *PI*43
> "*...... the proposition can be true or false can say no more than that we only predicate 'true' and 'false' of what we call a proposition.*" *PI*136
> "*....we understand the meaning of the word when we hear or say it......what we grasp in this way is surely something different from the 'use', which is extended in time!*"*PI*138

A language game is a form of life. Linguistic communication is inherently public, temporal, interactive, and built on shared conventions, which are in turn built around a social consensus. Language can be used for many different things, and there is no unassailable definition of a language game. One might wish to define a game, in general, as "a set of agreed conventions within an interaction between people." But even if such conventions are written down and stated as the game's definition, they may still evolve. Language games are inherently local, contingent upon a specific context that is necessarily narrowly defined. Language games have an explicit temporal component because they evolve and emerge over time, just as human life does. Language specifically, is a game that humans play, humans use language in a way that other forms of life do not: human language requires human thought, and human interaction. Superfluous rules may be discarded or forgotten, new rules may be added, each because participants of the game tacitly agree that such modifications are necessary. But "necessary" means "necessary in order to get on with the game", which presupposes some "essence" of the game that is above and beyond the rules as stated. Wittgenstein's point is that trying to formally pin down what this essence of any game is, is not possible, at least whilst using language.

> "*But how many kinds of different sentences are there? .....There are countless kinds.....And this multiplicity is not something fixed, given once for all;*

> ***but new types of language, new language-games, come into existence, and others become obsolete and are forgotten....Here the term 'language game' is meant to bring into prominence the fact that the speaking of language is part of an activity, or a form of life."*** *PI*23

The problematic nature of a "private language." A private language is a highly problematic concept, because the "speaker" of such a language would be unable to adequately determine whether its rules were being correctly followed. Following rules is, Wittgenstein argues, central to meaningful language. Rules are, by definition, social conventions that are agreed upon (explicitly or implicitly) by the participants of a language game. Suppose a person independently developed a set of linguistic rules within their own minds, shared these rules with no other person, and "spoke to themselves" internally in a private language that accorded to these rules. That person would (presumably) think that they were correctly following the rules of their private language. But there would be no set of external criteria by which such a claim could be assessed, therefore they would be unable to tell whether their belief was mistaken. This lack of an external reference point would also prevent the "speaker" of a private language realizing that they were "correctly" following the rules, but only doing so accidentally after having "misunderstood" them. The key point is that the very concept of having a language with rules breaks down when it is applied in a restricted manner within one individual's subjectivity. This brings into Wittgenstein's thought the fundamental question as to whether or not it is possible to think what one cannot express in language. Linguistic rules necessitate criteria by which the correct and incorrect application of these rules can be determined. Correct and incorrect application of the rules of language can only be meaningfully ascertained publicly - because "correct" means "makes sense and is useful to the other participant of the language game."

> ***"And hence also 'obeying a rule' is a practice. And to think one is obeying a rule is not to obey a rule. Hence it is not possible to obey a rule 'privately': otherwise thinking one was obeying a rule would be the same thing as obeying it."*** *PI*202

Evolution of meaning. Meaning is fluid and evolves over time, because use is extended over time. There can be no fixed definition of a language game, just a collection of fluid, evolving properties. A word is interpreted by a participant in a language game, at a point in time, and deemed to have a meaning, a use in that game. As a result other words and or actions follow. But the public nature of the mapping between a word and its use/meaning is intrinsically connected with the fact that the meaning can change over time, as the public consensus on which it is based moves along. This evolution of language games is reflected by the fact that various forms of meaningful language exhibit "family resemblances", i.e. varying degrees of overlap in the possession of multiple different properties, rather than a single definitive property. The closest one might get to such a form for meaningful language might be along the lines of "$mind \underset{\longleftarrow}{\overset{language}{\longrightarrow}} mind$" (with the arrows denoting interaction). But this is not so much a propositional form of what language is as a description of what language does. Exceptions can be found for any more detailed logical exposition of supposedly universal features of meaningful language. (Even gesturing, or the empathetic "reading" of a person's facial expressions count as linguistic communication). What matters is not the logical perfection, but getting one's point across in a context - and what this actually entails evolves over time.

> "*Instead of producing something common to all that we call language, I am saying that.....they are related to one another in many different ways.....these relationships.....we call them all 'language'.*" *PI*65
> "*How should we explain to someone what a game is? ....We do not draw the boundaries because none have been drawn....*" *PI*69

Subjectivity aids contextual understanding. There is a role for a subjective, intuitive "tuning in" to the "contextually objective" meaning of words, which is a natural part of language games. Subjectivity is an integral part of using language, and of interpreting its rules, in a "normal" contextual way, not something that can or should be eliminated from meaningful language. Meaning is public, and there is therefore no "hidden" meaning inside the ego trying to escape inner subjectivity. Subjectivity is required in order to fill in the gaps left by fluidity, redundancy and ambiguity in the use of words. Even if the grammar of language has fixed rules, within the relation *context* → *statement* → *meaning* there is still a significant role for subjectively, pragmatically, "knowing what the other person means" in a particular context. Interestingly, the fact that this kind of "reading between the lines" is a core part of meaning within everyday communication, combined with Wittgenstein's suggestions that subjective patterns convey meaning (an empathetic reading of facial expressions, or a work of art), strongly suggests that it is possible for us to "think what we cannot say." There a natural continuity between "inner" subjective content and "outer" linguistic expression of that subjectivity. The distinction between the two, and the Cartesian notion that inner subjectivity can be taken as a primary given from which all else can be derived, is artificially simplistic. The very notion of the distinction between an ultimate inner ego, which has a "state" that it is outwardly expressing in language, is a concept for which there is no need. Language is an extension of a subjective state, rather than an expression of it in some derived, objectified form. There is no inner meaning of words within the ego, which real words in real language games grasp at but ultimately fail to express comprehensively. Inner subjectivity is related to public use, but its relationship to this use is as fluid as any other element of our language games. This is reflected by the fact that language modifies subjective content - for example Wittgenstein notes that although animals think, there exist types of thought that humans have, as a result of their capacity for language, which cannot possibly be imagined for non-human animals.

> "*...A child has hurt himself and he cries; and then adults talk to him and teach him exclamations and, later, sentences..... the verbal expression of pain replaces crying and does not describe it.*" *PI*244
> "*One can imagine an animal angry, frightened, unhappy, happy, startled. But hopeful? And why not? .....the phenomena of hope are modes of this complicated form of life.*" *PIIIi*

Remarks on the foundations of mathematics: Foundationless useful conventions

Wittgenstein's extension of his "meaning is use" vision of language to mathematics, mankind's most useful language, was the piece of work that he himself regarded as his most significant[22,23,24]. However, the book's validity and significance is academically controversial. This is probably due to the fact that Wittgenstein's dismissal of Kurt Godel's famous "incompleteness theorems"[25] is widely held to reflect a misunderstanding of Godel's work. As one might expect based on Wittgenstein's other work, he seems to regard the very notion of a foundation for mathematics as bordering on meaningless, and think that if mathematics does have a foundation it is nothing more than utility. The strongest theme that emerges from *Remarks on the Foundations of Mathematics* is demystification. Wittgenstein wants to bring mathematics back down to Earth, re-ground mathematical concepts in the situations for which they were initially developed. and (as in his other works) sweep away the meaningless pseudo-problems that result from pushing the concepts too far from this grounding. It is perhaps tempting to label Wittgenstein's views as "intuitionism" (the idea that mathematical entities exist only in the minds of those doing mathematics). But some of what Wittgenstein says does not sit so easily with such a pragmatic view. He acknowledges that entities (like, say π) turn up unexpectedly in the course of mathematical investigation - and by implication are (arguably) therefore in some sense are closer to discoveries than inventions. But one imagines that he would have described the discoveries of mathematics as being discoveries in the world, moulded around the language of mathematics, as opposed to discoveries literally within mathematics in itself. He would, in any case, presumably have resisted classification into any such "school." The following general, inter-linked themes can be identified:

Demystification and contingency. Mathematics is (nothing more than) the application of useful rules, mathematical meaning is nothing more than the use of these rules, and mathematical meaning-objects have no independent existence outside of the minds of mathematicians. Proofs of theorems are accepted by virtue of their demonstrating the valid application of new rules in a particular context, or of existing rules in a new context. Accepting a proof amounts to a decision to adhere to such an application of a rule, as opposed to acceptance of the existence of some metaphysical truth. Mathematical "truths", if they can be said to exist at all, are a combination of tautological internal consistencies, and the useful application of such consistencies to the world. This amounts to a complete rejection of what one might call the "language of God" Platonist view of mathematics, in which mathematical entities have some independent, privileged or "transcendental" existence in their own right. This pragmatic view does not appear to raise a contradiction in Wittgenstein's mind with the synthetic (i.e. discovery like) *a priori* (i.e. experience independent) nature of some mathematical results. For instance, he explicitly describes how the distribution of prime numbers within the real numbers may be the best candidate for Kant's synthetic *a priori* knowledge (i.e. because this distribution is not inherently within the concept of a prime number). The notion of "rule following" is prior to that of the validity of mathematical proof, and separate from it. Wittgenstein's primary aim seems to be to demystify "puffed up" mathematical concepts and unwarranted extrapolations from them. He argues that the notion of construction of a new proof blends in to the concept of applying accepted rules in a new way, with no clear-cut distinction between the two (and no need for one). Mathematicians don't discover anything in mathematics, they invent new rules - and that is all that needs to be said.

> "*The mathematician is an inventor, not a discoverer.*" *RFMI*168
> "*The distribution of primes would be an ideal example of what could be called a synthetic a priori.... it is at any rate not discoverable by analysis of the concept of a prime number.*" *RFMAIV*43

Irreducibility of mathematics to logic. Mathematics is not reducible to logic, because logical expression of mathematical ideas renders those ideas non-mathematical, therefore useless, therefore meaningless. In directly disputing the notion that mathematics is reducible to logic, Wittgenstein dismisses the possibility of some sort of universally applicable logical calculus. Reducing mathematics to logic inherently involves representing mathematical relationships in a non-mathematical way (e.g. describing them as functions of subordinate functions in a hierarchical way, and, in so doing, inappropriately averaging over the details of the functions themselves, removing their capacity for mathematical work). This "logicization" of mathematical relationships produces an end result that is useless, therefore meaningless. Wittgenstein additionally contrasts the certainty that can be achieved within a proof derived from geometric representability, from one derived from deductive inferences - casting doubt over whether the two methods of proof are necessarily analogous. Aswell as losing mathematical power, attempts to encapsulate mathematics in logic lead, if they are to be invertible, to the false assumption that mathematical functions can be assigned arbitrarily - i.e. that anything can be described mathematically. This leads to pseudo-mathematical/logical relationships being defined that are too vague to be useful, and consequently not really mathematics. Crucially, Wittgenstein does not reject the possibility of identifying a degree of correspondence between mathematics and logic. Rather he attacks the hierarchical notion that all mathematics is reducible to logical/deductive rules. If a logical representation of a particular piece of mathematics is possible, the reason for any correspondence between logic and mathematics cannot be attributed to one or other language. For instance, (to take a crude example) the truth of the statement that the ratio between the circumference and diameter of a circle is π, is not something that there is any logical reason for, it is merely an "empirical" observation. It seems that Wittgenstein views the idea of putting logic at the foundation of mathematics as implicitly equivalent to attempting to reduce such geometrical facts to syllogistic reasoning. The key point that Wittgenstein reiterates is that even though it might be possible to do this (i.e to imagine a series of logical arguments about "why" $\pi = \frac{circumference}{diameter}$), if such arguments were posed in logical form (as opposed to just being a restatement of the rules of geometry), they would no longer be fit for geometric use, therefore no longer be meaningful. If one wanted to calculate the circumference of a circle, one would want elementary geometry. It would not be possible to "rebuild" geometry from an esoteric, artificial set of logical arguments supposedly underlying "why" geometry is the case. We should forget such useless attempts at esoteric abstraction, and just use the mathematics.

> "*Nothing is more likely than that the verbal expression of the result of a mathematical proof is calculated to delude us with a myth.*" $RFMAIII25-26$
> "*Generality in mathematics does not stand to particularity in mathematics in the same way as the general to the particular elsewhere.....one can know a proof thoroughly....... and yet at the same time not understand what it was that was proved. ......Now when does one understand it? I believe: when one can apply it.*" $RFMAV25$

Groundlessness. Wittgenstein does not seem to think any defensible metaphysical foundation for mathematics exists, and appears to regard the search for such foundations as deeply misguided. In this sense, a core aspect of Wittgenstein's philosophy of mathematics (like many of his other works) is represented by the idea of groundlessness[25]. Mathematics works, and particular sub-systems of mathematics are internally consistent; they provide a "firm grounding" when we work within them. But explanations as to why they work cannot be provided by the systems themselves: the ground is itself groundless, and any attempt to rectify this is based on a fundamental misunderstanding of the meaningfulness of language. Mathematical concepts give the impression that we must dig deeper so as to find the foundation of such grounding- but that such digging leaves us back where we begin, suggesting we should stop digging and simply use rules in the world. The search for the "foundations" of any system of thought cannot be comprehensible within that system of thought, let alone form part of that system. This is arguably similar in spirit to Godel's incompleteness theorems - suggesting perhaps that Wittgenstein's objection to these ideas was based more upon what he saw as a misuse of mathematical formalism than a difference in perspective. Again, utility is the only basis by which mathematics is "inexorable", and the only foundation that mathematics (or any other language) needs.

> "*What does mathematics need a foundation for? It no more needs one, I believe, than propositions about physical objects - or about sense impressions, need an analysis. What mathematical propositions do stand in need of is clarification of their grammar...*" $RFMAVII17$

A limited, pragmatic role for subjectivity exists in the useful application of mathematical rules. Acceptance of a proof or mathematical proposition is, if genuine and warranted, connected to a compulsion to accept that proof as agreeing with what "makes sense" relative to both explicit, and, to a limited extent intuitive, criteria. There is a difference between this pragmatism and trusting one's individual subjectivity unwisely, e.g. over the consensus agreement that results from many different people performing the same mathematical operations. Subjective acceptance of some rules as being useful, some not. Subjective acceptance of implicit inferences about the axioms on which mathematical statements are based (i.e. with incomplete knowledge of these axioms) is part of the real, useful, public consensus of doing mathematics. Acceptance of this public consensus above (say) one's own "feeling" for the answer to a mathematical question is also, ultimately, a subjective judgment. There is also a (much more important) role for subjectivity at the point of acceptance (or not) of what constitutes a new proof - i.e. when it is reasonable to accept that there has been some form of modification to grammatical/mathematical rules, and that this modification is satisfactory. The more academically desirable idea of complete "foundational" knowledge (of every formal axiom and related concept), before accepting a proof or result, is more explicitly consistent than the "everyday" acceptance of mathematical rules, and lacks the need for a subjective idea of what "makes sense." But ultimately even this sort of academic abstraction still runs up against the groundlessness of mathematical concepts, which is something that must simply be accepted. There is a role for subjective intuition, not in repetitively doing established mathematics, but in deciding what kinds of mathematical question to ask, and (in particular) not to ask. This reflects a general principle: Public language use is "objective" and verifiable, only relative to conventions. But what these conventions actually are and why we adopt them is something that necessarily brushes up against the boundary of subjectivity.

> "*Whoever calls '$\neg\neg p = p$' (or again '$\neg\neg p \equiv p$') a 'necessary proposition of logic' (not a stipulation about the method of presentation that we adopt) also has a tendency to say that this proposition proceeds from the meaning of negation. When double negation is used as negation in some dialect, as in 'he found nothing nowhere', we are inclined to say: really that would mean that he found something everywhere. Let us consider what this 'really' means.*" $RFMAI11$

The nonsensical nature of the results of applying mathematical operations to inappropriate contexts. Extrapolation from mathematical concepts, beyond a frame of reference in which these concepts are useful, is unwarranted and problematic. In the full spirit of the *Tractatus*, Wittgenstein attacks what he sees as the pseudo-metaphysical meaningless language resulting from attempts to apply mathematical concepts well beyond the frame of reference to which they are suited for use. Attempts to provide "foundations" for certain concepts when no such foundations are possible, leads to inappropriate and unnecessary attempts to make entirely different ideas consistent with one another. Wittgenstein rejects such attempts as meaningless and obviously silly: he particularly targets "incompleteness theorems" and certain applications of infinity. Wittgenstein does not directly engage with the theorems that result from the confusions he discusses, he simply hints that the very assumptions and preliminary arguments on which the discussion of such theorems is based are ridiculous. Two specific examples illustrate Wittgenstein's perspective:

So-called "incompleteness theorems" are confused and problematic, due to the nonsensical nature of both the notion of "completeness", and that of defining propositions self-referentially in such a way that they assert their own unprovability. The conceptual basis of Kurt Godel's first incompleteness theorem is the idea that no theory capable of describing elementary arithmetical operations can be both consistent and complete. "Theory" in Godel's sense, means a collection of sentences in formal language, adhering to semantic rules. "Consistent" means that none of the rules within a theory contradict one another. "Complete" means that all semantically valid statements within the set of rules of the theory are provable within that theory. Godel's theorem[24] is of the form "Say that theory T contains proposition P, and that the meaning of proposition P within theory T is "P is unprovable." Then P cannot be proved within T. If P is provable, T entails P's provability, and is not therefore *consistent* with P's definition (as unprovable in T). If P is unprovable, theory T contains a proposition P that cannot be proven within it, therefore T is not *complete*. Incompleteness theorems discuss the issues surrounding the fact that the concepts of completeness and consistency are incompatible. But incompleteness theorems rely on a general initial statement of the form "Suppose that P says that it itself cannot be proven. Now, suppose P is true....." Wittgenstein argues that the very form of such statements is ridiculous. He directly disputes the very idea that it is allowable to postulate a proposition that implies its own unprovability. He appears to consider it silly to perceive the problems arising from such borderline self contradictory statements as real. This is primarily, it seems, because statements that contradict themselves could never be expected to be useful. In the *Tractatus* Wittgenstein dismissed Russell's paradox, largely on the basis that one cannot simply assign set membership arbitrarily (sets and the things in them must be different kinds of things). Similarly he dismisses incompleteness theorems on the basis that a proposition's meaning cannot just be decided arbitrarily; propositions cannot simply be defined as contradicting their own provability in the very context in which they are defined, and still treated as legitimate. In any theory there are inbuilt constraints upon what that theory can do (this is actually in direct agreement with the basic "spirit" of the incompleteness theorems, which is presumably why Wittgenstein's criticism of Godel was not taken seriously). But Wittgenstein seems to consider the very idea that there is any conversation to be had about this matter as silly. Crucially, for Wittgenstein, in saying that something is unprovable, one loses the right to speak of it as being true or false, and any paradox arising from the nature of the relationship between its truth/falsehood and its provability simply disappears. One simply cannot say things like "P is unprovable, suppose P is false...." (or true), nor build one's thinking around such statements.

> "*The proposition 'P is unprovable' has a different sense afterward - from before it was proved. If it is proved, then it is the terminal pattern in the proof of unprovability. - If it is unproved, then what is to count as a criterion of its truth is not yet clear, and-we can say-its sense is still veiled.*" *RFMII*16
>
> "*.....the proposition asserting of itself that it is unprovable is to be conceived of as a mathematical assertion.........when we said 'it asserts of itself'- this has to be understood in a special way.......Godel's proposition, which asserts something about itself, does not mention itself.*" *RFMAVII*21

Countability and infinity are mutually exclusive. Some properties are attributable to finite, countable collections of entities, and other properties are attributable to theoretical collections of entities resulting from infinite repetition of a certain operation. Ideas, such as Cantor's continuum hypothesis, which presuppose that both types of properties are, or should be, equivalent or relatable to each other, are confused to the point of being meaningless. Wittgenstein dislikes arguments that mix up the concepts of countability and infinity. The "Continuum hypothesis"[26] states that there is no set with more elements than the (infinite) set of all natural numbers (i.e. the set of all positive and negative integers), but less elements than the (also infinite, but necessarily larger) set of all real numbers (i.e. the natural, rational and irrational numbers). The issue as to whether the continuum hypothesis can be proven or not (it cannot be proven in conventional set theory[26]) is relevant to modern discussions about the "foundations" of mathematicians. Cantor's[26,27] (unsuccessful) attempts at a proof of the continuum hypothesis were based on what has become known as his diagonal argument - which shows that there are infinite sets that cannot be put into one to one correspondence with the natural numbers. Although the set of all natural numbers, and that of all real numbers are both infinite, the latter is necessarily larger than the former. Based on the diagonal argument, one cannot assign a cardinality (determine reliably the number of elements in the set) to any set intermediate in size between these two sets. Therefore one cannot prove the continuum hypothesis to be false. The diagonal argument is based on the idea of arranging the (infinite) set of all real numbers in decimal form as a matrix. If one could do this, one could assign a cardinality to a set somewhere between the set of all natural numbers and the set of all real numbers. The diagonal argument shows that the infinite divisibility of the number line means that there will always be a number left out of any such matrix, hence the set of all real numbers cannot be counted. For instance, consider the set of all possible infinite sequences of binary digits, e.g.

$$\begin{array}{l} S_1 = (0 \quad 0 \quad 0 \quad 0 \quad 0) \\ S_2 = (1 \quad 0 \quad 0 \quad 0 \quad 0) \\ S_3 = (0 \quad 1 \quad 0 \quad 0 \quad 0) \\ S_4 = (0 \quad 0 \quad 1 \quad 1 \quad 0) \\ S_5 = (1 \quad 1 \quad 1 \quad 1 \quad 1) \end{array}$$

, and so on, in a collection of infinite size, up to S_n where $n \longrightarrow \infty$. Given such an arrangement, one can always start at the top left corner and work down the diagonal, successively pick the opposite digit to the one present at that position in the matrix, and create a set that differs from each set in the matrix at at least one position (e.g. for the matrix just given one could write $S_6 = (1, 1, 1, 0, 0)$). The same principle (of always being able to pick a different digit along the diagonal) was used by Cantor to show that the collection of decimals that would result from collecting all the real numbers could never be countable - preventing disproof of the continuum hypothesis. In Wittgenstein's view, this theme has given rise to pseudo-philosophical concepts relating to the supposed existence of two different types of infinity, and to misguided questions as to whether sets can be perceived as "the basis of" numbers. Wittgenstein implies that the very notion of the provability (or not) of the continuum hypothesis is mistaken - because of a fundamental disconnect between the concept of cardinal number (the number of elements within set) and real number. The number of elements of a set (i.e. "the cardinal numbers" as a concept) is necessarily a positive integer or zero, due to the binary true/false nature of belonging. An element either is in a set or is not, if there is any sort of continuum between the two, the very concept of an element's belonging to a set breaks down. When speaking about "the real numbers" as a concept, one must deal with negative, non-integer, irrational numbers, etc - and there is no reason to think these should be compatible with extrapolation from the concept of cardinality. The idea that one should be able to prove the Continuum hypothesis presupposes the idea that one can count a collection that is by definition uncountable, something that Wittgenstein sees as ridiculous. He appears to wish to demystify the whole discussion, by stating that if a concept involves infinity, it is simply absurd to discuss it in the same sentence as countability or 1 : 1 ordering. For Wittgenstein, the concept of natural number is a very practical idea, the concept of infinity is far less tangible, and the idea of mixing up the concept of order within a countable collection, with the concept of infinity, is laughable.

> "*The dangerous, deceptive thing about the idea: 'The real numbers cannot be arranged in series', or again 'The set ....is not denumerable' is that it makes the determination of a concept.....look like a fact of nature.*" *RFMII*19
>
> "*one pretends to compare the 'set' of real numbers in magnitude with that of the cardinal numbers.... The difference in kind between the two conceptions is represented, by a skew form of expression, as difference of extension....I believe, and hope, that a future generation will laugh at this hocus pocus.*" *RFMII*22
>
> " *'Ought the word 'infinite' to be avoided in mathematics?' Yes: where it appears to confer a meaning on the calculus: instead of getting one from it.*" *RFMII*58

The tension between temporality and rigour in language. The more unassailable a language is, the less temporal that language is, and mathematics is the least temporal, therefore most unassailable, human language. A natural tension can be imagined between a language with mutable rules, the conventions of which evolve over time (as does its meaning) versus a temporally fixed language. In other words there is a natural tension between a given language's evolvability and its consistency. Mathematics is more powerful than other human languages due to two properties. First, its practical utility (no other language can be used to do engineering), second its consistency over time (the geometry and arithmetic used today is largely identical to that used by the ancient Greeks). It is natural to postulate that these two properties are connected - the practical power of mathematics goes hand in hand with the fact that it is fixed in time. (New mathematics is, of course, produced, but in general constitutes more of an addition to, than an alteration of, the mathematics that already exists). Wittgenstein does not go into great detail on this subject or make this argument explicitly - but he does hint at the connection. The decision to accept a rule and take it as a given in a language game, as opposed to go through the process of identifying the need for that rule in the game, is related to acknowledging the consistent applicability of that rule. This amounts to giving a rule a timelessness. Wittgenstein discusses the analogy of the number of sounds a word produces when we say it. One might argue that this basic principle is consistent with the fact that mathematics loses some of its explanatory power in academic subjects in which time, the memory within systems, and temporal changeability, is more relevant. For instance, the mathematics used in Newtonian physics is fully deterministic, whereas the application of mathematics to subjects such as biology or economics is both stochastic (and in general less powerful in terms of its capacity to explain).

> *"When we say 'This proposition follows from that one' here again 'to follow' is being used non-temporally."* *RFMI*103

Although the three works above are Wittgenstein's most coherent and most respected, additional posthumous publications have been developed based on various collections of his notes. In most of these writings the style is closer to *Philosophical Investigations* than the *Tractatus* - I.e. a list of problems, unanswered questions and thought experiments, designed to suggest, rather than explicitly assert, the argument to the reader. The main themes in these works are reiterations of those at the core of Wittgenstein's thought; that meaning is use and there is no need for it to be anything more, that the usefulness of meaningful language is just part of our form of life, that linguistic clarification allows us to sweep away unanswerable pseudo-problems, and that passing questions over in silence is something significant and important. Over and above these basic ideas, a few additional points can be identified, mostly relating to the application of Wittgenstein's basic ideas to traditional metaphysical questions and to the philosophy of mind.

Zettel: Mind, body and language

Zettel[27] is a collection of Wittgenstein's notes published by Anscombe and Von Wright in 1968, taken from Wittgenstein's notes between 1929 and 1948. He emphasizes questions relevant to consciousness, mind and matter, and the relationship between thought and language.

The unfakeable, future-directed will. Intention (i.e. "the will") is the most unfakeable, and most private aspect of the human form of life, and is directed towards the future in a manner that distinguishes it from other aspects of our subjectivity. Intention involves a connection to the object of intent that is not present for other conscious states. This is because intention is a forward-looking, rather than a discrete thought or "state" of consciousness that can be pinned down or divided into constituents. Importantly Wittgenstein implies that this generality, this holism, results in intention containing an incommunicable aspect. Similar arguments apply to expectation - which is a similar general outlook, which doesn't necessarily have to lead to the manifestation of the thing being expected. The inherent involvement of our consciousness with the future is something that we can neither fake nor comprehensively communicate. Linguistic expressions of the will can suit practical purposes, and generate appropriate empathetic responses in others, but such expressions are fundamentally incomplete and superficial. The clearest way to explain this is perhaps to note that it is not possible to genuinely make someone "want" something (in the way that it is possible to "make" someone perform some physical action on command). Nor is it possibly to reliably know for certain the "wants" of another person.

That this statement is true, despite the fact that it is a natural and commonplace part of our language games to express intentions and desires, seems to be the key distinction that Wittgenstein is trying to get at. Although all aspects of consciousness are to some extent private, the future-directedness of intention, gives it a particular in-communicability. This is also inherently connected to the fact that what what intends need not be the same as what works out as happening. This special privacy makes it unfakeable. We are our will, therefore it is impossible to ever know or express what it might be like to be a human and lack the property we refer to as free-will. From this perspective, questions like "How do I know I have free will?" become perfect examples of linguistically meaningless pseudoproblems that can be neither asked nor answered -because being the manifestation of (apparent) intent is a prerequisite for asking such questions. One cannot involuntarily will something, consciousness and will are terms that, to the extent to which they actually mean something, are essentially interchangeable: we are our intention/will/consciousness, and this fact underlies any question that we may ask, or experience that we may have.

> "*Compare these orders: 'Raise your arm. Imagine... Work...out in your head. Consider...Concentrate your attention on...' with these: 'Intend...Mean by these words...Suspect that this is the case. Believe that it is so. Be of the firm conviction...Remember that this happened...Doubt whether it has happened...Hope for his return.*'" $Z51$
> "*... how do I know this movement was voluntary? - I don't know this, I manifest it.*" $Z595-600$

Intuitive, Empathetic recognition of consciousness. Consciousness is temporal, indivisible, and recognizable only through instinctive empathetic pattern recognition - through which its presence in others is as clear as in ourselves. Consciousness has an indivisibility to it that makes it impossible to reliably establish a relationship between states of consciousness and time. For similar reasons it is not possible to differentiate between states of consciousness and innate dispositions. Thus the individuality and temporality of consciousness are connected. When it comes to conscious interaction, the eye, nominally a physiological organ of sensory perception, provides a means by which to be interactive with other consciousnesses, as opposed to the "receiver" of a "signal" from the world. Seeing and looking are concepts that necessarily entail a degree of empathetic recognition and conscious interaction, in addition to just the detection of physical pattern through a physiological interaction with light. In other words we "see with our consciousness" aswell as with our eyes. Empathy is a pre-linguistic (conscious) pattern recognition instinct, which cannot be pinned down in language. Subjective empathetic recognition of another consciousness is pre-conceptual (therefore incompletely communicable), and is connected to the wholeness of consciousness. Because of the strength of this empathetic recognition, many professional philosophers have argued[3] that Wittgenstein dismisses the question as to the existence of "other minds", in the sense that we recognize consciousness in others just as easily as in ourselves, thus "proof" of the existence of our minds is no more (or less) of a problem than proof of our own consciousness. Wittgenstein would presumably dismiss both problems as part of the unknowable grounding for any human life or language. Were "I" to doubt that I have consciousness, I would necessarily doubt my own doubt, because there is no "I" without my consciousness. If I doubt my own consciousness, there is not only nothing that I cannot doubt, but nothing that I can say either. Doubting the existence of the consciousness of others is, for Wittgenstein, ridiculous, because I instinctively recognize it as well as I recognize anything. The "question" as to the existence of consciousness in both myself and others, is within a the category of issues that, if doubted, necessitate the effective abandonment of any kind of coherent thought.

> " *Look into someone else's face, and see the consciousness in it and a particular shade of consciousness....The light in other people's faces. Do you look into yourself in order to recognize the fury is his face? It is there as clearly as in your own breast.*" Z220 – 221
> " *We do not see the human eye as a receiver, it appears not to let anything in, but to send something out. The ear receives; the eye looks......One can terrify with one's eyes, not with one's ear or nose...*" Z222 – 223

Irreducibility of consciousness. There is no reason to expect consciousness to be explicable in terms of brain activity. Conscious sensory perception is temporally emergent, generative and self-referential. There is no reason to suppose consciousness to be reducible to what might be called "causal atoms", (i.e. physical first-causes, supposedly on some epistemological level distinct from consciousness, such as the memory trace or eve the gene). There is mind, there is matter, they are associated. But "causation" between mind and matter (in either direction) is a completely misguided way to think about this association. Wittgenstein seems to regard "mind-body problems" or attempts to "localize" consciousness as further examples of pseudo-philosophical meaningless language. He speaks as if the subjective experience of consciousness, and the material of the brain, are two obviously correlated, but fundamentally different, processes, and one cannot say that one "causes" the other. The expectation of a systematic relationship between "objects" of consciousness and "objects" in the world results in confusion, because there are no such separable "objects" within either level of explanation, let alone some mapping between the two levels. (In fairness to reductionists, however, Wittgenstein does not, of course, suggest how a productive scientific methodology could be built from his way of looking at things. Nor does he address the traditional conception of the mind-body problem, i.e. how mind and body can influence each other if they are fundamentally different). Wittgenstein appears to take it as indisputable that mind and body are different, and different types of explanation are warranted for each. Thus, attempts to ask "But what exactly is...?" of a particular subjective state is an attempt to muddle two distinct kinds of explanation. Emotions are conscious states with expression behaviour, which may sometimes be directed at objects in the world. They are not localized or (neurologically) causally reducible. This non-localizability distinguishes emotions and other conscious states from physiological sensations, which Wittgenstein appears to view as distinct from authentic conscious phenomena (and presumably neurologically reducible to a far greater extent). One imagines Wittgenstein might say that the definitive difference between a conscious state and a physiological response is that a conscious state exhibits some form of connection with intention, with the indivisibility of consciousness and the sense in which it is "directed at" the future, whereas a physiological response is "just" a property of the body, with no more connection to intention than any other feature of the environment the mind might react to. (However this is not a distinction Wittgenstein makes clear). Wittgenstein seems critical of any notion of applying causation to the phenomena of mind, including a system of mental causation that works in parallel with that of the physiological, which he seems to regard as a variant of the perspective that views the brain as "causing" conscious, and one that implies the existence of the very "ghostly soul nature" that it is trying to expel. He gives the example of a temporally "generative", i.e. self-referential, temporal feedback dependent, connection between a seed and a plant, as the closest analogy to the connection between mind and brain. He suggests that in the plant-seed relation, aswell as the mind-matter one, expecting a reducible 1 : 1 mapping at any isolated point in time is a misunderstanding. Saying what a conscious state "is" entails separating that state from consciousness, which simply cannot be done. One gets the impression that Wittgenstein simply does not think any notion of causation is relevant to consciousness - because it such a temporal, indivisible, and foundational phenomenon.

> "*The general differentiation of all states of consciousness from dispositions seems to me to be one that one cannot ascertain by spot-check whether they are still going on.*" Z72
> "*One no more feels sorrow in one's body than one feels seeing in one's eyes.*" Z495
> "*No supposition seems to me more natural than that there is no process in the brain correlated with associating or with thinking; so that it would be impossible to read off thought-processes from brain-processes. I mean this: if I talk or write there is, I assume, a system of impulses going out from my brain and correlated with my spoken or written thoughts. But why should the system continue further in the direction of the centre? Why should this order not proceed, so to speak, out of chaos?........a seed always produces a plant of the same kind as that from which it was produced- but nothing in the seed corresponds to the plant which comes from it; so that it is impossible to infer the properties or structure of the plant from those of the seed that it comes out of- this can only be done from the history of the seed...*" Z608 – 609

Temporal evolution of meaningful language. Language games have a natural temporal evolution and fluidity of meaning, connected to an inseparability from their context, and from the feedback that exists between thought and language. It is therefore misguided to expect a single permanently meaningful language to be possible. Meaningful language is the interaction between an innate predisposition to pattern recognition (which is analogous to a physiological response), thought, and the modification of thought via public language games. Our language moulds our thought over time; animals' lack of language is both a cause and an effect of their lack of human thought. Thought can sometimes surpass language (Wittgenstein mentions poetry and music). Thus, language continually needs to "catch up" with thought. This, in turn, necessitates an indeterminacy, i.e. a mutability, within language. Wittgenstein uses colour as an example of the feedback by which language informs and shapes concepts - pointing out, for instance that although recognition of colour may be innate to our capacity for sensory perception, the "concept" of colour requires development through a language game. Implicitly therefore, an emergence occurs in our interpretation of words and statements in language, that contextualizes the meaning of the words we hear; a kind of mental "tuning in." Additionally, Wittgenstein reiterates his argument that the proper role of philosophy is to "cure" the philosophical impulse. By "philosophical impulse" he appears to be referring to a pre-disposition for an individual to get wrapped in a feedback between thought and language that lacks any meaningful or "useful" tie to the world. Wittgenstein uses the phrase "slow cure" to describe this contextualization and correction of language misuse, and connects it to mathematicians being (in his view) bad philosophers, due to the "bitty" discretized thinking that mathematics requires. The indefiniteness and fluidity, which would be erroneous in mathematics, is necessary for the natural temporal evolution of other human languages.

> "*We now have a theory, a 'dynamic theory' of the proposition; of language, but it does not present itself to us as a theory. For it is the characteristic thing about such a theory that it looks at a special clearly intuitive case and says: 'That shows how things are in every case; this case is the exemplar case of all cases'- 'Of course! It has to be like that' we say, and are satisfied. We have arrived at a form of expression that strikes us as obvious. But it is as if we had now seen something lying beneath the surface....*"
> Z444

On Certainty: The relativity of knowledge to the possibility of doubt

Wittgenstein's last work, *On Certainty*[28] was written in response to G.E. Moore's "*A defence of common sense*"[29]. Moore's paper, as the title suggests, was a reaction against the kind of radical skepticism that Wittgenstein espoused, and is built around certain basic claims that Moore argues he knows to be quite obviously true of his own relationship to the world. Moore further argues that every other person has an equivalent set of claims, which they, equivalently, also know to be true for themselves and their basic relationship to the world. Moore's intention appears to be to point out that there are certain claims that are so foundational that it is dishonest and silly to claim to doubt them. This kind of pragmatism is, Moore argues, one and the same with certainty, in so far as this word means anything. Moore asserts that he has a body with certain physical properties relative to an external physical world, a world that had existed for many years prior to his own existence, and that he has an awareness of the existence of other people:

> "*(1) I begin, then with my list of truisms, every one of which (in my own opinion), I know, with certainty, to be true......There exists a living body, which is my body. This body was born at a certain time in the past.....there have....been large numbers of other human bodies......the earth had existed for many years before my body was born.......I have....had many different experiences......in the case of very many other human bodies who have lived upon the earth, each has been the body of another human being..*"

He goes on to claim that many other people can claim an analogous set of things to be true:

> "*(2)... human beings who have had......experiences of each of the kind mentioned in (1)...each has frequently....known.......a proposition corresponding to each of the propositions in (1)....*"

Here Moore seems to be appealing to a basic empathetic sense of common ground with others, and to the authority of a common sense consensus. Moore makes an arbitrarily long list of "common sense" facts, implicitly held by everyone during the course of practical life, regarding their body, and the physical manifestation of their consciousness. Moore says that he "knows" that all of the facts on this list are true, and that other human beings have a directly comparable list that they, analagously, also know to be true. This exercise appears to be an attempt to bring back down to Earth the concept of what it is to "know" anything. Moore's purpose in doing this appears to be to criticize what he sees as the absurd position of philosophers (most explicitly Wittgenstein) who claim to doubt such basic common sense knowledge:

> "*They seem to think that the question 'Do you believe that the earth has existed for many years past?' is the sort of question which can be properly met by: 'It all depends what you mean by 'the earth' and 'exists' and 'years" .....It seems to me that such a view is as profoundly mistaken as any view can be.*"

The sort of knowledge that Moore identifies in his truisms can be divided into the basic spatial/temporal and conscious knowledge of oneself (truism 1), and the empathetic/consensus-driven claim that (almost) everyone else has an equivalent list of common sense knowledge (truism 2). He then distinguishes philosophers who claim that this kind of physical/spatio-temporal and empathetic knowledge is untrue, from philosophers who acknowledge that it is true, but state that it is only true in a common sense way, which not sufficient for us to claim to know it to be "philosophically true" or "metaphysically certain", a category that can be differentiated from that of "true by virtue of common sense." Moore disagrees with philosophers in both categories, for different reasons:

> *".....many philosophers have really held views incompatible with (2) ... may, I think, be divided into two groups: A. What (2) asserts is.... at least partially false..... B...with regard to any of the classes in (2).....that no human being has ever known, with certainty, that any propositions of the class in question are true."*

Moore then goes through a process of attempting to refute these two categories of philosophical positions that doubt common sense knowledge, by pointing out what he views as contradictions in the basis for this doubt. His arguments amount to the idea that philosophers who doubt common sense use common sense in order to express this doubt. For Moore, philosophers in group A, those who doubt the truth of common sense physical facts or conscious awareness (truism (1)), and/or doubt that there are other human beings who have come to the realization of such facts (truism (2)), deny their own existence in having such doubt. By denying their own existence they deny their own doubt, rendering the whole dialogue through which such doubt might be expressed a nonsense, stating (with reference to the class of propositions (2) above): *"If .....no proposition of that class is true, then no philosopher has ever existed, therefore none can ever have held.... that no proposition belonging to it is true.."* Moore argues out that doubting the existing of others belies a hypocrisy in philosophers who refer to the existence of other philosophers, because they need to doubt the existence of these other philosophers in order to be consistent. (Moore does not appear to permit such philosophers the ability to doubt their own existence and still make arguments, or indeed to make arguments whilst doubting those arguments). Furthermore, Moore also accuses those philosophers who allow his common sense assertions (1) and (2) to be stated on a pragmatic level, and consider them true in terms of common sense, but doubt that they are known with "certainty" to be guilty of a dishonest inconsistency:

> *"If he says 'These beliefs are beliefs of common sense, but they are not matters of knowledge', he.... is actually asserting the existence of many human beings ......They seem to me to......regard the proposition that those beliefs are beliefs of common sense, or the proposition that they themselves are not the only members of the human race, as not merely true, but certainly true...."*

I.e. Moore takes issue with the claim that philosophers accept "common sense truth" of some sort, but deny that common sense truth is philosophical certainty. He (it seems) says that philosophers who are sure of certain things in a common sense way, simply are certain of these things. The idea of some distinction between common sense and philosophical certainty are dishonest and unnecessary. Thus, Moore argues that philosophers who doubt common sense are contradicting themselves, and philosophers who do not doubt the common sense itself, but doubt its certainty, are being dishonest. In the rest of his essay Moore talks about the implications of the basic common sense knowledge that he defends. Moore seems to think that there are facts of consciousness that may be about the present, the past, or "timeless." He defends the idea of common sense facts about the world, stating that his two initial truisms "*...are propositions which, in a certain way, imply the reality of material things, and the reality of space.*" as well as that "*Time is real, and (in a certain sense) at least one self is real.*" Moore decides that there are both mental and physical facts (though he never defines a fact, apparently taking it as another example of self-evident common sense knowledge). Again, he then goes through a process of accusing those who doubt such facts of hypocrisy. Interestingly, Moore does brush up against the notion of groundlessness, stating that "*there is no good reason to suppose either that every physical fact is logically dependent upon some mental fact or that every physical fact is causally dependent on some mental fact*", and (most especially) that with regard to certain propositions of common sense knowledge "*I...while holding that there is no doubt whatever that many such propositions are wholly true, hold also that no philosopher, hitherto, has succeeded in suggesting an analysis of them....which comes anywhere near to being certainly true.*" In this second quotation Moore is arguing that we lack the means to appropriately analyze common sense knowledge, even though we know that it is true. (He does not appear to notice that this implies two different senses in which one can "know" things, therefore implying both that a sense of knowing things that would undermine common sense is at least conceivable, and that the kind of distinction between philosophical certainty and common sense that he so fiercely criticizes is at least consciously real, for him aswell as the philosophers he is criticizing).

The key consideration that Moore neglects is the need for an explicit and consistent relationship between words and criteria by which we decide whether or not words mean anything. Every "fact", every "truism" is conditional to a meaning criterion. Philosophers can, if they wish, doubt any knowledge of their own existence, and that of others, whilst still "knowing" common sense things- because, implicitly or explicitly, they have different criteria for designating "knowledge" in each case. In relating the words to the knowledge criteria we end up, of course, with what Wittgenstein called a language game. Some such games may assign meaning according to practical utility, others according to some sense of satisfaction based on perceived metaphysical coherence. But the key point is that one needs to define the meaning criteria before launching in to the discussion. Because Moore makes no distinction between common sense and metaphysical/academic "philosophical knowledge", numerous problems arise in his arguments, mostly relating to his apparent faith in consensus. The list of "common sense facts" that some individual people would include in their idiosyncratic truism (1) would differ from the facts that others might put on their own personal list of common sense knowledge. (To pick the obvious example, a religious person might add a creation event, an atheist might include planetary accretion - the point being that the inclusion of "historical facts" raises subjective judgment calls). This is related to various semantic problems, such as defining a "fact" and deciding what it means to "know" something. If one assumes that there is no robust and unconditional definition of fact, knowledge, or anything else, then Moore's essay reduces to a series of almost entirely unjustified assertions of "things everybody just knows." Who exactly is supposed to adjudicate the boundaries of this democratically defined category of common sense knowledge is not something Moore makes clear. It seems that Moore (with obviously humble and honourable intentions) seeks to critique any separation between pragmatic and academic/philosophical "knowledge." He does not think philosophers should be allowed to claim to doubt things that they clearly implicitly assume to be true in their everyday lives (e.g. the reality of other people or of the physical world). One is not allowed, in Moore's view, to make pragmatic assertions (or act upon them) that one still doubts. He would presumably have been uneasy with the notion of "levels" of knowledge, or the necessity of contextualization of linguistic statements to a particular set of rules. He also believes that there are certain things that everyone "just knows" to be true, and his humility makes him critical of any "lofty" attempts to doubt such basic things within the isolation of academic philosophy: "*...to speak with contempt of ..."Common sense beliefs" which I have mentioned is...the height of absurdity.*" At times one gets the impression that Moore would only accept the authenticity of a philosopher's doubt if that doubt was so crippling that the philosopher in question was afraid to move or interact with anybody. Most glaringly, Moore never defines the boundary delimiting common sense, beyond which legitimate and non-hypocritical doubt would become permissible.

Unsurprisingly, Moore's faith in common sense was not something shared by Wittgenstein. This difference is related to a different level of concern for, and attention to, the meaning of words between the two men. In *On Certainty* Wittgenstein points out the various implicit assumptions in Moore's arguments, demonstrating the relativity of any form of knowledge or linguistic expression to a grounding - one that is itself ultimately impossible to justify. The spirit of Wittgenstein's book is made most clear by the first line. Moore had discussed the relationship between knowledge and sense data by describing various interpretations of what one is entitled to claim to "know" when holding one's hand up in front of one's face, for instance regarding the relationship between the sense data received by the mind and the object itself. Examples of this description include: "*This is part of the surface of a human hand*" ("This is my hand" is apparently not allowed because we do not see "a" hand, only one side of it), or

> "*There is some relation R, such that...'There is one thing and only one thing, of which it is true both that it is part of the surface of a human hand, and that it has R to this sense datum'.....*' "

Wittgenstein doesn't disagree with the analysis, merely the confidence that Moore (and other empiricists) place in it. He begins with the (slightly punchier) line "*If you do know that 'here is one hand' we'll grant you all the rest....*" Three inter-related and fundamental arguments can be identified in *On Certainty*:

Certainty means that doubt is useless, not that doubt is absent. Drawing the boundary that defines where doubt becomes useless is a subjective act of assertion, not a deductive act relative to some set of rules. This is because any such deduction would require the existence of a universal frame of reference in which doubt/certainty could be decided, and no such frame of reference exists. To say, as G.E. Moore does, that one is "certain" of a particular collection of "facts" by appealing to "common sense", amounts to subjectively asserting that it is of no use to doubt such "facts." This entails an equivalent implicit assertion as to what is meant by "usefulness", "practical value" or some other such perceived quality - by which the use (or lack thereof) of doubting a given fact can be assessed. Any given justification for a particular choice of "value criteria" as to what one assumes to be beyond reasonable doubt, itself requires justification; "usefulness" must be defined relative to a perceived purpose, which in turn must be justified, leading to an infinite regress that has to end with a subjective choice of value criterion. For example, if a person holds their hand in front of their eyes and says "I am certain that this is my hand", a second person might ask them "But how do you know that you can trust what you see with your eyes?" The first person might reply "I just know, it is ridiculous to doubt what I see right before my own eyes." The frame of reference within which one might doubt what is before one's own eyes, wonder whether one's own sense data is an illusion, etc, is inconsistent with the first more "common sense" perspective. But any criteria by which one perspective might be favoured over the other would ultimately be a subjective choice of the point at which one says "it is ridiculous to doubt such things." Crucially, Wittgenstein argues that when Moore says he is "certain" of the truisms that he discusses, what he is really saying is that he considers it useless (or otherwise not to his taste) to doubt the assumptions upon which those truisms are based. Wittgenstein points out that in practice, what we call "certainty" amounts to this sort of subjective assertion of the uselessness of doubt. Moore's manuscript is, in this sense, his personal subjective assertion about the particular set of things that he considers it useless to doubt. Moore seeks to give authority to his subjective choice of doubt/certainty boundary by appealing to consensus; i.e. by picking facts that "everyone knows" under the umbrella of collective, public, reasonable common sense. Wittgenstein points out that, strictly speaking, there is no good reason for this subjective choice. If facts are chosen for the list about which there is less consensus this subjectivity becomes more apparent. Learning which doubts are useful and which doubts are not in a particular context constitutes learning a language game (but nothing more). Nevertheless, without taking some things as being beyond doubt, there is no possibility of participating in language games. The fundamental point is that the choice as to what one assumes to be beyond reasonable doubt, and what to be open to doubt, is both unavoidable and, ultimately, unjustifiable.

> "*If you do know that here is one hand, we'll grant you all the rest. When one says that such and such a proposition can't be proved, of course that does not mean that it can't be derived from other propositions; any proposition can be derived from other ones. But they may be no more certain than it is itself..... What we can ask is whether it makes sense to doubt it.*" $OC1-2$
> "*The reasonable man does not have certain doubts.*" $OC220$
> "*What would it be like to doubt now whether I have two hands? Why can't I imagine it at all? What would I believe if I didn't believe that? So far I have no system at all within which this doubt might exist.*" $OC247$
> "*Why doesn't Moore produce as one of the things that he knows, for example, that in such-and-such a part of England there is a village called so-and-so? In other words: why doesn't he mention a fact that is known to him and not to every one of us?*" $OC462$
> "*If Moore is attacking those who say that one cannot really know such a thing, he can't do it by assuring them that he knows this and that.....Moore's mistake lies in this-countering the assertion that one cannot know it, by saying 'I do know it.'.* " $OC520-521$

Universal epistemological relativity. Both doubt and knowledge are, without exception, relative to claim-specific assumptions. The claim "I know x with certainty" means only "If I were to doubt x then I would have to consider everything else that I think I know open to doubt.", or (equivalently) "I am as sure of x as I am of any other element of my knowledge.": Certainty is merely the assertion of one's (subjective) lack of doubt. Philosophical discussion of claims of certainty clears up meaningless language by demonstrating this unavoidable relativity to foundational assumptions. Practical "common sense" assertions such as those made by Moore are not sufficiently relativized to explicit assumptions to be meaningful. Moore does not recognize and/or make explicit the assumptions that he is implicitly making in trusting his common sense truisms; in essence he merely asserts these truisms without "doing philosophy on them", and as a result he never defines a set of criteria by which one can decide on the meaning of "knowing" something. By refusing to make this separation between common sense, practical "knowledge" and explicitly defined relativity between statements in language and (groundless but explicit) assumptions, Moore makes it impossible to establish any criteria for the meaning of his claims. Wittgenstein's intention seems to be to point out that there is a difference between saying "doubting x is silly and of no practical value" and saying "I know x with certainty." This is the essence of the difference between the two men's perspectives. Saying "I am certain of...." is an attempt to instill on someone else a subjective conviction (of the useless of a particular doubt), whilst pretending one has a logical argument. A coherent/meaningful logical argument would need to be of the form "In a context in which set of assumptions A_x can be held true, x is true." With claims as basic as the above "I know I see my hand before my eyes" (assuming one is holding one's hand before one's eyes), a logical form of the claim would be "If I have cause to doubt the truthfulness of this being my hand before my eyes as I hold it up, there is nothing that I do not have cause to doubt, because the assumption that I can trust my immediate sense data is a prerequisite for everything else I consider myself to know." Doubt, if it is to be posed meaningfully, must also be made relative to assumptions. One cannot say "I doubt everything", because one would have to doubt that one doubted- i.e. without knowledge there is no doubt. One can only doubt something because it is inconsistent with something else (a meaningful statement would be "assuming A_x, I doubt the truth of x", the key point again being relativity). Wittgenstein's task is to remind Moore, and other defenders of common sense, of an underlying subjectivity in their choice of assumptions, and force them to make these assumptions clear.

> "*If you tried to doubt everything you would not get as far as doubting anything. The game of doubting itself presupposes certainty.*" $OC115$
> "*The child learns by believing the adult. Doubt comes after belief.*" $OC160$
> "*To say of man, in Moore's sense, that he knows something; that what he says is therefore unconditionally the truth, seems wrong to me. - It is the truth only inasmuch as it is an unmoving foundation of his language games. I want to say: It's not that on some points men know the truth with perfect certainty. No: perfect certainty is only a matter of their attitude.*" $OC403-404$
> "*A doubt that doubted everything would not be a doubt.*" $OC450$

Time and epistemological faith. Time creates serious problems for claims of "certain" knowledge (even when they made are appropriately relative to the assumptions that they rely upon). An underlying faith that the future will be like the past is necessary for any language game to work coherently. The relationship between assumptions, and the propositions made relative to those assumptions, evolves naturally over time. This necessitates that the language game relevant to the particular "assumption$\longrightarrow$proposition" relationship evolves too. The process of evolution moves the subject of the language game away from the game itself, meaning that the game must be corrected, must "catch up." The validity of inductive reasoning is, ultimately, only an assumption, and tomorrow the world could change so as to show us that the assumption$\longrightarrow$proposition relationships we are confident in today either no longer hold, or were never valid in the first place. The assumption that this will not happen is just that, an assumption. What is perhaps a more serious extrapolation from the line of thinking that Wittgenstein goes down here, is that the possibility of there being ways in which the future is qualitatively different from the past or present entirely undermines the premise of certainty. Designation of the boundary between what is and is not usefully open to doubt, and thus what is and is not useful to assume as certain, itself also changes over time. This fluidity is made particularly apparent by the obvious historical evolution of what people have considered to be certain, some examples of which Wittgenstein mentions. The key point is that any given language game works (produces usefulness, meaning, relative knowledge) at a point in time, but the notion that this will remain the case in the future is a matter of belief.

> "*The more primitive idea is that the earth never had a beginning. No child has reason to ask himself how long the earth has existed, because all change takes place upon it. If what is called the earth really came into existence at some time -which is hard enough to picture- then one naturally assumes the beginning as having been an inconceivably long time ago.*" *OC*182
> "*If someone doubted whether the earth had existed a hundred years ago, I should not understand, for this reason: I would not know what such a person would still allow to be counted as evidence and what not.*" *OC*231
> "*We say we know that water boils and does not freeze under such-and-such circumstances. Is it conceivable that we are wrong?.......Whatever may happen in the future, however water may behave in the future, - we know that up to now it has behaved thus in innumerable instances. This fact is fused into the foundations of our language games.*" *OC*558

Heidegger

Being and Time[30] is widely regarded as Martin Heidegger's definitive work. It is an exploration of human subjectivity that is rich, verbose, deeply confusing and willfully cryptic - but at the same time exceptionally perceptive and insightful. A key question that Heidegger felt had been lost since the beginnings of philosophy was: "What is the meaning of being?" He never clearly answered this question, but he argued that it was unavoidable to ask it, because the question itself connects deeply to the core of what we are. Heidegger does seem to have had the feeling that a least part of the answer is temporality. That is, a question that he was unable to answer, and that we understand intuitively but are unable to express adequately, is connected to the fact that our subjectivity itself is temporal, that we are, in some sense, embodied time. The skeptical reader might wish to note that Heidegger's long, dense, semi-incomprehensible, interconnected prose, not to mention the vague and gratuitous vocabulary that he invents, would certainly fail the "meaningless language test" of analytic philosophy. In this vein, Bertrand Russell described Heidegger's work as:

> *"Highly eccentric in its terminology, his philosophy is extremely obscure. One cannot help suspecting that language is here running riot. An interesting point in his speculations is the insistence that nothingness is something positive. As with much else in existentialism, this is a psychological observation made to pass for logic."*[31]

There is a good deal of legitimacy to Russell's skepticism. His point about the "positivity" of nothingness presumably relates to how Heidegger advises us to, in various ways "just be" in the world and confront the finite nature of our existence. But, crucially, Heidegger's starting point was the intuitively important subjectivity that Wittgenstein felt compelled to recognize but pass over in silence.

Heidegger's core themes are authenticity, time, an inescapable connectedness and interactivity, and the revealing/disclosure of truth. We have a pre-conceptual understanding of the inescapability of our connection to the world and our interactivity with it. This understanding is closely related to an intuitive appreciation that there is something behind the concept of Being, i.e. the meaning of being, that we are compelled to ask about (but which we can't precisely define). This preconceptual "sense", which is pointed at by the notion of the meaning of being, distinguishes us (i.e. human beings), from the other things in the world (which we can divide into background objects and tools for use). Temporality is the backbone of what we are, and is linked to our predisposition to subjectivize time in a historical manner, relative to events that are important to us. We are also in time, i.e. we experience a constant underlying continuity between birth and death, which grounds us, inescapably and irreducibly, into the world. The fundamental temporality of our subjectivity is connected to an awareness of our inevitable death. Fear of death causes anxiety, and this anxiety is important. We can respond to this anxiety, aswell as to other features of our embeddedness in time (like the call of our conscience through guilt) via an outlook that is authentic or one that is inauthentic. In essence, authenticity involves confronting the truth directly and placing ourselves in the context of the connectedness between our birth and our death, by looking back to the past and forward to the future, and seeing the links between the two honestly. Inauthenticity involves an immersion in the present, and in some instances an active hiding from this underlying temporality, so as to duck the anxiety we face in confronting our fear of death, and our reluctance to be receptive to the truth. Inauthentic human being in the world is characterized by a curious, chattery, uncertain collective insanity that Heidegger calls the they-self. The they-self is a kind of "everyone and no-one" persona that assists the individual human being in hiding from the future and immersing itself in the present, by distraction through gossip and superficial trivial curiosity. This is easy to allow ourselves to become immersed in, but within ourselves we are ultimately aware that in doing so we are merely hiding from our fear of death. If we wish, and depending upon the situation, we can "snap out of it" by confronting our anxiety and guilt and extracting ourselves from the they-self, and remembering who we really are, in what Heidegger calls a "moment of vision." When we respond authentically to a moment of vision, we recover a real, truthful outlook that looks back historically, both to our individual birth and to a wider historical heritage, and forward to our own death, anxious about its inevitability, but nonetheless with an honest and unafraid outlook. This forward-looking, authentic outlook in which we individualize ourselves from the they-self and directly confront the future relative to our fear of death, Heidegger calls anticipatory resoluteness. As well as being "more real" relative to our true selves, this outlook makes us receptive to the disclosure of truth, i.e the unveiling of a previously hidden, underlying consistency. This disclosure of truth Heidegger calls "aletheia." Achieving this authenticity by being aware of our temporality and mortality, is the closest thing we have to an "answer" to the question of the meaning of being, i.e. the closest conceptual approximation to the essence of what it means to be a human in the world.

Being and Time

In more Heidegger's key concepts within his definitive work are as follows:

Being. A foundational, inescapable, irreducibility that is a prerequisite for all metaphysics. Humans have an intuitive pre-conceptual understanding of Being, but it can only be defined tautologically relative to the being of particular things. We cannot really answer questions about "the meaning of Being" in the way that we can answer more precisely defined questions, but we have an intuitive awareness that such questions are nevertheless fundamental. Being (which it is tempting to define crudely as "everything that is, and how it is") is an unbounded whole that cannot be grounded in other concepts or defined with respect to them, but is rather the basic interactivity upon which such things rest. The various senses in which we might use the word "is" today are: (*i*) The predicative- something "is" some descriptive property, i.e. what that something is, (*ii*) the existential, something exists, i.e. that it is, and (*iii*) equivalence, i.e. the identity focused way in which we say something is the same as something else. In contrast to these senses, Heidegger focuses on "how-being", as distinct from the more conventional "that-being" (existential) or "what-being" (predicative)[4]. Heidegger notes that being is self evident, and defines it in an unapologetically circular manner, such that it cannot be defined with reference to something else. This lack of grounding may have resulted from Heidegger's wish to distance himself from a hierarchical categorization of being (i.e. use of the word "is") dating back to Aristotle. In this earlier tradition, various entities (objects, people, quantities, relations) were defined as having a being relative to something else - leading to an infinite regress in which ultimately the being of anything is defined with respect to a fundamental category of "substances" (which are also problematic to define). Heidegger asks, from an explicitly human point of view, exactly what the being is of something that would ask questions about that being. He asks, how are we (or any other entities) in the world? He divides everything that is into what is "present-at-hand" (objects), what is "ready-to-hand" (tools and equipment) and "daesin" (human being). The final concept, daesin, is Heidegger's central one. And it is daesin that has the intuitive, pre-conceptual awareness of the question of the meaning of Being. All of these concepts are collected under the umbrella term "Being."

> "*.....all ontology.....remains blind and perverted from its ownmost aim, if it has not first adequately clarified the meaning of Being, and conceived this clarification as its fundamental task.*" $BT10-11$

Presence-at-hand versus readiness-to-hand. Two distinct ways of perceiving the things we encounter in the world, distinguished by the manner in which we anticipate that those things relate to our intentions. Crudely speaking, things that are present at hand are "mere" things, relatively inconspicuous inanimate objects that simply blend into the background (e.g. a pebble on a path that we walk over without noticing). Things that are "ready to hand" are tools or equipment that we perceive as being potentially useful for a tangible purpose, something through which we can imagine ourselves interacting with the world (e.g. a tool or piece of equipment that we associate with a specific task). The classic example discussed with reference to this distinction is a hammer. A child or a person unfamiliar with principles of construction might perceive the hammer as present-at-hand, as a mere stick of wood with metal attached to one end. But anyone who actually knew what a hammer was would view it as ready-to-hand, as a potential tool with which one might hammer in nails so as to build something or fulfill some sort of goal. A hammer handle in which the metal head was lost would be merely present at hand as a piece of wood. Thus, readiness-to-hand involves an inherently more purposeful, interactive perspective that Heidegger argues is more natural to us. Importantly, to see what is present at hand, one has to view these things in a context that may not be obvious to us, because it is not connected to our being, which is focused on purposeful interactivity. For example, to see the wood in the handle of a hammer as composed of lignin (wood) macromolecules, one has to analyze that handle biochemically, from the perspective of atomic theory, biochemistry, biology, etc. This perspective is far removed from the simple, proximate availability of the hammer as a tool, by which we merely use the hammer without thinking about its macromolecular constituents. For Heidegger, the way in which we normally encounter things in the world is by virtue of their being ready to hand, whereas seeing things as present at hand is a derivative and lesser ("deficient") way of seeing those things. Detached, theoretical circumspection, in which we may perceive the world as just things that are present at hand are, for Heidegger, ways in which we step away from authentic human-ness. We have intentions, for which we use equipment to do things, to such an extent that when equipment use is done well, we may forget we are using a tool at all (a good builder, absorbed in his work, may feel as if the hammer he is using is an extension of his own hand). This co-ordinated equipment use is, for Heidegger, closer to what we really are as human beings (than, say, viewing the hammer as a stick of wood with a metal add-on, or a heterogeneous collection of molecules). It is worth noting that the distinguishing qualities that differentiate the ready to hand perspective from the present at hand, (usefulness as a tool, the capacity to interact with the world, purposefulness), are fundamentally temporal. The concept of interaction does not work for an isolated moment in time, one interacts with the world through an object that is ready to hand for a purpose, over a time interval.

> "*.....we come up against unusable things within what is ready-to-hand already....it seems to lose its character of readiness-to-hand. It reveals itselfs as something just present-at-hand and no more.....*" *BT*73

Daesin. The entity for which the question of the meaning of being has significance, through its pre-conceptual understanding of that question. Daesin is human being-in-the-world, capable of asking questions of itself about the meaning of that being. The word translates from the German "da", meaning "there", or "here", depending on the context, and "sein", meaning "being" or "to be." So the direct English translation of Daesin is "being-(t)here." Daesin is more temporal than it is spatial. It experiences anxiety, conscience, moments of vision, moods, an awareness of others - all of which it can encounter through the adoption of various "modes of being" (some of which Heidegger finds favourable, some not - the latter he labels "deficient" modes of being). Various derivative concepts can be discussed in relation to Daesin, such as care, modes of being, moods, authenticity and resoluteness.

> "*Daesin is an entity for which, in its Being, that Being is an issue.*" *BT*191

Care. A prerequisite psychological/emotional disposition to perceive, and develop emotional attachments to, aspects of the world. Care forms the fundamental basis of Daesin, human being in the world, and in turn the basis of care is temporality. Heidegger suggests that people have an inescapable predisposition to caring, which is what sets Daesin apart from other categories of being. Thus, temporality is the meaning of the being of care; Daesin is grounded in care, care is grounded in time. The capacity to be concerned about things, to will things, like things, experience emotional moods etc, goes hand in hand with a tendency to look forward to the future. The temporal quality of this core stems from how Daesin, because it has care, is ahead of itself - i.e. there is something in our subjectivity that cannot help but look to the future in a subjective manner. Through care, Daesin is always ahead of itself, even whilst it is alongside objects or whatever else it encounters in the world. Care is circular - the basis of Daesin runs ahead of itself, projecting itself into the future and looking back to the past. This circularity, Heidegger argues, is what makes the tautological character of many of his definitions unavoidable and still meaningful.

> *"We have defined Daesin's Being as 'care'. The ontological meaning of 'care' is temporality."* BT364

Thrownness. An intuitive appreciation of the contingency and helplessness of how Daesin exists in a world to which it is unavoidably connected. Thrownness can be interpreted as a awareness of a lack of control over our fate and our being in the world. Human beings are thrown into the world (they did not enter the world by choice) and what happens to them is often outside their control, and indeed inconsistent with what they want. We experience "factical" (contingent, tangible) elements of the world, and these things are temporal in character, and we go along with them. The term "facticity" is ambiguous, but refers to both literal observational facts about experience, and an underlying intuitive sense that such concrete facts are contingent, a reminder of how we are thrown into the world. There is something anonymous about our thrownness, which means that when we notice it, we are more alone with ourselves than in other more usual situations in which we forget about it. This difference in appreciation of thrownness occurs via the different moods we have (i.e. is intuitive more than it is rational), the most important mood of which is anxiety. Thrownness is also conceptually related to the fact that (in Heidegger's view) it is impossible to prove the existence of an external world (and rather silly to attempt such a proof). This inseparability between ourselves and the external world explains why Heidegger favoured a phenomenological philosophical method in which experiences and consciousness are studied at face value as what they are, rather than being framed in terms of subsidiary, smaller phenomena on which they are presumed to depend. There is an "uncanniness" to our existence in a contingent world, an intuitive feel that there is something intrinsically improbable about our existence, some sense in which we (i.e. human beings) do not quite fit comfortably with the world, but this feeling evades coherent description. Thrownness is something which we intuitively pick up through moods, and which we often dismiss (by hiding in the mentality of an anonymous they-self), because of the unpleasant anxious feelings that awareness of thrownness elicits within us. Confronting this contingency in our existence is a fundamental part of attaining authenticity.

> *"As something thrown, Daesin has been thrown into existence. It exists as an entity which has to be as it is and as it can be....."* BT276

They-self. An anonymous, collective, and inauthentic group persona, in which individual Daesin becomes lost in the being with others, ultimately as a way of hiding from its fear of death. The they-self refers to a kind of herd mentality in which individuals do what conforms to the social expectations to which they are subjected. Becoming lost in the they-self amounts to doing things because that is "what one does", or thinking things because "that is something everyone knows", (as opposed to doing things and thinking things because of some form of clearer individual judgment or conviction). Heidegger uses the phrase "Das man" when he discusses the they self, for which a direct English translation is "the they." When discussing the they-self Heidegger refers to our everyday interactions with other Daesin via a collective mentality, characterized by "idle talk" and gossip, insubstantial and shallow curiosity, and a kind of falling by which we lose our individual authenticity by following the crowd, as well as an ambiguity or duplicity within the meaning of the subject matter being chattered about. This collective mentality penalizes and negatively judges individuals who go against it, for instance by gossiping in a malicious manner. This is the norm - most of our existence is spent lost in the they-self, recovery of individual authenticity is rare and often unpleasant. If we want to discover ourselves, our own individual being, and do so in an authentic manner, we need to break away from the they-self, which acts as a barrier to this discovery and this breaking away. When we do so we lose the comfort of feeling an affinity to the they-self, and experience anxiety, but for Heidegger such anxiety is the price we need to pay for being in the world in an authentic, genuinely individual manner. This act of authentically breaking away from the they-self and experiencing this anxiety is closely related to a fear of death, to confronting the finite nature of our existence.

> "*Anticipation reveals to Daesin its lostness in the they-self, and brings it face to face with the possibility of being itself..........released from the illusions of the 'they', and which is factical, certain of itself, and anxious.*" *BT*266

Fallenness and curiosity. The manner in which Daesin loses its true self to the immediate present, through the idle talk, curiosity and ambiguity that is associated with the they-self. In fallenness Daesin becomes absorbed in the distractions of superficial interaction with others through this they-self (as opposed to any detached contemplation or focus on the future). In reference to fallenness, Heidegger mentions novelty, but in a near derogatory sense, as a shallow and empty curiosity that is related to idle chatter. Curiosity, the German word for which, Neugier, literally means "lust for novelty", is not conceived of by Heidegger as metaphysically significant, but rather is a way of jumping from one thing that has not thus far been experienced to another, as a means of distracting oneself (from one's true self), rather than to understand the things the curiosity is directed at. Distraction within this shallow curiosity is an example of falling - the manner in which authentic Daesin loses its true self in the world, i.e. falls into an inauthentic mode of being. Fallenness thus results from this uninquisitive, shallow curiosity, idle, gossip-like chatter, and a duplicity of meaning that results from this sort of chatter.

> "*Fallenness'......' means an absorbtion in Being-with-one-another..... guided by idle talk, curiosity and ambiguity.*" *BT*175 − 176

Anxiety. An uncomfortable but authentic state of emotional discomfort, through which we intuitively but directly confront our fear of death and the finite nature of our existence. Moods are temporal in nature and reveal an authenticity to us, and anxiety is the most important mood. Anxiety manifests as a "objectless angst"[4], and is the result of our intuition rebelling against the inauthentic activities that we get caught up with, as a result of our thrownness and contingency of our existence in the world. Anxiety picks up on the "uncanniness" of this thrownness and contingency ("facticity"). Anxiety is closely related to fear, which might be considered an unproductive counterpart of, or reaction to, anxiety. We react with a cowardly fear and inauthentic incoherence when we do not respond to anxiety in a way that is "loyal" to our true selves. Anxiety is an emotional response to a remembering of the likelihood of a bad future, the likelihood of the repetition ("being-as-having-been") of a bad past. This is connected to our natural lack of comfort with thrownness, and the fact that we all too easily forget this discomfort (in a they-self or otherwise), rather than confront it. Anxiety is what happens when we do not forget these disconcerting features of our existence. Aswell as bringing us face to face with the possible of a bad future, anxiety simultaneously brings us face to face with the alternative possibility - an authentic future, in which we confront our thrownness. To discover such authenticity, we must be receptive to anxiety. In other words, for Heidegger, anxiety, far from being a bad thing, is a fundamentally important feature of our being, something by which (if we respond to it appropriately) we can recover the essence of ourselves.

> "*....anxiety as a basic state of mind belongs to such a self understanding of Daesin, on the basis of Daesin itself. Being-towards-death is essentially anxiety. This is attested..... by Being towards death.......when it perverts anxiety into cowardly fear and, in surmounting this fear, only makes known its own cowardliness in the face of anxiety.*" *BT*266

Conscience. The call of care. An inner, intuitive reminder of the existence of authenticity and the primordial knowledge we have of it (roughly speaking an "inner voice" reminding us of how, as we already know, we are doing something inauthentic, untrue to what we are). Conscience operates through feelings such as anxiety and guilt. Anxiety is different from fear because it is more vague, it lacks a specific referent - we just generally feel anxious. Guilt is a similarly general emotion, we may feel guilty about particular debts we owe, but Heidegger distinguishes this more common use of the word from a generic, underlying feeling of guilt - we feel guilty that we are not confronting what we really are and what our being in the world really constitutes. In this second type of guilt we do not owe a tangible debt to another, we feel guilty "for" ourselves because we are not acting authentically. Traditionally the idea of a conscience relates to some form of inner voice that assesses particular actions as good or bad according to certain criteria. In Heidegger's concept of conscience, actions are assessed not through social ethical-moral value systems, but rather an inner, fundamentally individual subjectivity, which has a primordial knowledge. Conscience doesn't tell us what to do, but it does tell us when we're doing the wrong thing. Conscience is both generated and received by individual Daesin, and is the label Heidegger creates for this inner monologue concerning authenticity. Conscience causes us to feel an underlying anxiety and guilt, and is the means by which subjective realizations of inauthenticity are reached. Daesin may be lost in the (inauthentic) they-self, and may be woken from this kind of state by its Conscience. The capacity to listen to one's conscience is the result of a choice to free oneself from the they-self. This decision to become receptive to conscience is a kind of "coming into its own" for Daesin.

> "*Conscience manifests itself as the call of care: The caller is Daesin, which....is anxious about its potentiality for Being. The one to whom the appeal is made is this very same Daesin, summoned to its ownmost potentiality-for-Being.*" *BT*278

Authenticity. A coming to terms with the world in a truthful and receptive way, as a consequence of free choice, by individual Daesin, closely related to an awareness of the inevitability of death. Authentic Daesin makes the free choice to confront its own ultimate mortality by (for instance), extracting itself from the tranquilizing distraction of the they-self. Authentic Daesin has "won" itself, it is its own, knows itself, and is open to the truth. Inauthentic Daesin is, by contrast, distracted, detached from itself, and on some level hiding from death's inevitability. Heidegger states that inauthenticity is not a "lesser" mode of being (although the emphasis he places on attaining authenticity seems to suggest otherwise). Inauthenticity is a commonplace mode of being, arguably the norm - work must be done by the self in order to live authentically, and authenticity may never occur for a particular individual, the may remain lost in inauthenticity throughout their existence. Nevertheless, authenticity is Daesin's original state; we get lost in inauthentic ways of being when we are thrown in to the world. Authenticity relates to an open-ness, a receptiveness, to the subjective call of conscience and to the wider disclosure of the truth. An authentic perspective allows an individual to mentally traverse their whole existence, placing their being in context, and adopting a perspective of receptiveness to the truth.

> "*authentic disclosedness shows the phenomenon of primordial truth in the mode of authenticity, The most primordial, and indeed the most authentic, disclosedness....is the truth of existence.*"*BT*221

Death. The certainty of the finite nature of an individual Daesin's existence, that "pulls" this existence along and underlies it. Death is the end of an individual Daesin, and is a fundamentally individual and non-relational experience. When we die we truly alone, and we can only genuinely be authentic if we confront this fact and modify our outlook accordingly. Anxiety relates to a subconscious fear of death, and authenticity is closely related to confronting this fear. An awareness of death is close to an awareness of the fact that it is theoretically possible that nothing need ever have existed at all. An appreciation of this, combined with one of death's inevitability, creates an appreciation of a continuity between birth and death that colours and underlies human subjectivity, distinguishing birth and death from other "mere" events. Daesin is a connection between birth and death (as opposed to being something that happens to occur between these events). The looking forward to the future and back to the past, with birth and death as reference points, is what distinguishes human being, its subjectivity, and its capacity to occur in inauthentic or authentic modes of being, from other categories of existence.

> "*Death does not just 'belong' to one's own Daesin....Death lays claim to it as an individual Daesin.*" *BT*263

Disclosedness. An open-ness within Daesin to a process of revealing that happens in the world. Heidegger discusses various examples by which disclosedness permits a change in perspective within Daesin, generally one that is to a more truthful or authentic. Disclosure, (i.e. un-closure, the removal of some form of "closing") is a telling choice of wording, in the sense that it implies a layer of hidden-ness that is removed - either by the change in attitude within Daesin or by the change in the world. So roughly speaking, disclosure happens in the world, disclosedness is Daesin's receptiveness to this disclosure. This can be a subjective, subconscious process that can happen through moods and emotions. For example, Heidegger argues that anxiety discloses the world in its true state, through a kind of pro-conceptual, intuition-driven awareness, very different to a reason-based "working out" of the truth.

> "*Disclose' and 'disclosedness'........ shall signify 'to lay open' and the character of having been laid open.*" *BT*75

Resoluteness. Projection of Daesin on to its own Being-guilty. One experiences guilt, which is the call of one's conscience, "telling" oneself about inauthenticity, about one's being in the world in an inauthentic manner that is not loyal to oneself. Resoluteness is a decision to confront this guilt as what it is, i.e. to acknowledge this call of conscience, and to "project" oneself towards one's future possibilities in the light of this call. Resoluteness is therefore an outlook/perspective that results from a free choice. When one acts resolutely, one confronts the call of conscience, is disclosed/receptive to the truth, and is loyal to one's authentic self. For instance, a resolute person/Daesin thinks about the fact that they will die and that they fear this, and makes a conscious decision to consider this (as opposed to, for example, hiding in the anonymity and transient distraction of the they-self). Heidegger describes resoluteness as a "mode of Daesin's disclosedness", which (perhaps) might translate as a particular way of being open to revealing (of truth, of authenticity, of the implications of the call of conscience through guilt, etc). For Heidegger a resolute individual Daesin is a deeply and genuinely individual entity, and in choosing resoluteness it has recovered something that, prior to that choice, had been lost. Heidegger describes resoluteness as the authenticity of care. Thus, resoluteness is the authenticity of the basis of Daesin. In simpler non-Heideggerian language one might therefore translate resoluteness as "authentically being a human, in a receptive and forward-looking way."

> "*Resoluteness constitutes the loyalty of existence to its own self.....In resoluteness lies..... every possible moment of vision that may arise from it.*" *BT*391

Anticipatory resoluteness. Answering the call of conscience and using this to confront one's own mortality. Resoluteness combined with a directed mindfulness of the inevitability of death, and a consequent awareness of the potential possibilities for being in the world. Projection of Daesin on to its own Being-guilty (resoluteness), as a means of Being-towards death (anticipation), reveals its "ownmost potentiality" for Being (possibilities of what Daesin could do, that have been hidden). Anticipatory resoluteness is resoluteness coloured with an explicitly temporal outlook in which one is mindful of one's own mortality. Anticipatory resoluteness refers to Daesin adopting a resolute perspective and anticipating its own death, and as a result being mindful of the possibilities for its own Being in the world, and (above all) authentically an individual entity. Simplistically one might say that whilst resoluteness refers to acknowledging our conscience by confronting guilt, anticipatory resoluteness refers to acting upon this acknowledgment - so as to think about our own death and confront its inevitability. Anticipatory resoluteness is very close to a definitive example of authenticity. Resolute Daesin is authentically itself, and resolute Daesin that looks forward in the context of its own mortality is explicitly aware of death. This awareness is what truly "individualizes" Daesin, like no other feature of experience. So anticipatory resoluteness, like no other quality of human being in the world, is both authentic and individual.

> "*...resoluteness is constantly certain of death- in other words since it anticipates it- resoluteness thus attains a certainty that is authentic and whole.*" *BT*308

Moment of vision. An instance of revelation/unconcealing of the truth, combined with an arrival within Daesin at an understanding of that truth and its previous concealment. During a moment of vision, Daesin is fundamentally individualized because it has acknowledged the call of conscience and thereby been extracted from the they-self, so as to "be-towards-death" (appreciate its potential non-existence, and the potential impossibility of any existence), and in turn to be receptive to authenticity and the disclosure of truth. Crudely speaking therefore, the moment of vision is when an individual human being "snaps out of it" and experiences the revealing of an authentic core to their existence. The moment of vision is connected to a confronting of anxiety, an acknowledgment of its existence, and as a result of this acknowledgment, the development of a resolute outlook on future possibilities, in which Daesin's potential end leads to it taking a more authentic outlook on its future.

> "*That present which is held in authentic temporality and which thus is authentic itself, we call the 'moment of vision'........the moment of vision permits us to encounter for the first time what can be 'in a time'*" *BT*338

Aletheia. Disclosure/revelation of an underlying truth. The becoming non-hidden, specifically to Daesin, of something fundamental within the world. Heidegger's concept of truth combines an element of the more common notion of truth as consistency or convergence, with an element of revelation to human beings specifically: there would be no truth as associated with aletheia if there was no Daesin. So truth is something important that is revealed, specifically to Daesin, during instances of disclosure. It seems that Heidegger discusses truth as disclosure (as opposed to, for example) "becoming", because he wants to retain the kind of timeless foundational constancy underlying truth as traditionally conceived, but at the same time emphasize both disclosure and the human-specific, cognitive aspect of truth. After an act of disclosure, the truth that has been unconcealed moves into being ready to hand, i.e. being more consistent with truth as agreement. Disclosedness, because it is a key attribute of Daesin and its temporal nature, it what makes Daesin "in the truth", but derivative, inauthentic modes of being in the world cause Daesin to also at times be lost in concealed untruth. This combination of true things in the world, and truth as an attribute of Daesin's disclosedness, give a duality to Heidegger's thoughts on truth, within which aletheia is the act of revealing.

> " *the concept of aletheia.....the entities of which one is talking must be taken out of their hiddenness....they must be discovered. Similarly 'Being false' amounts to deceiving in the sense of covering up...*" *BT*33

Historicality. The predisposition within Daesin to attribute significance to past events, particularly when deciding what to do in the future. Historicality, Heidegger argues, is an outlook in Daesin that leans towards historiology, the study of history. Crudely speaking, Heidegger is saying that humans are naturally closer to historians than to, say, scientists. The way in which Daesin is in time, is toward its death, memorable of significant past events, etc, is Daesin's historicality. Historicality is related to the attitude/outlook of Daesin at the point of the moment of vision, relative to the possibilities available to it, to its fate of inevitable death, and to the heritage it carries with it from its past, towards this future fate. When authentic, historicality allows the repetition of the past to come together with resolute looking forward to the future. Historicality can also be inauthentic, absorbed in the history of the they-self, from which Daesin takes a "world history." But when authentic, historicality allows a resolute connection between birth and death, in an authentic, individual Daesin.

"Historizing" is, roughly, what Heidegger labels the way in which human beings relate themselves to their birth and their mortality, in time, how Daesin "stretches itself along." Authentic historicality is the kind of historizing that lies in anticipatory resoluteness, the kind of relating oneself to one's birth, one's past, one's heritage and historical nature, and one's inevitable death - through the forward looking (anticipatory), authentic, individual, truthful, mindful outlook of anticipatory resoluteness.

> " *fate - that is to say, authentic historicality...... The authentic repetition of a possibility of existence that has been....*" *BT*385

Temporality. The foundational cause of the indivisibility and connectivity of Daesin's existence. The subjective priority of time relative to other foundational elements of Daesin always ultimately amounts to the role of temporality in human existence, however this role manifests specifically. For instance, authenticity marks out Daesin that is more true to its own self than inauthentic Daesin, and a key difference between authenticity and inauthenticity is the manner in which authenticity stretches across time from birth to death, whereas inauthenticity is focused on the present (normally distracted by the short term trivialities of the they-self). Thus although Daesin is spatial as well as temporal, the title of Heidegger's book is not "Being and space" because his various themes (of opposing authenticity and inauthenticity, distraction and disclosure of truth, etc), all share a reliance on the human experience of time. Time permits historicality (looking back to attribute significance to past events), time is what allows Daesin to look back to its birth, to feel the pull of its death (therefore anxiety, therefore guilt and conscience, providing the potential to achieve authenticity) and so forth. Time is the fundamental "grounding" of Daesin, of human being. One could go as far as to say that the meaning of Being is temporal (although Heidegger does not say this directly). All of the above core concepts can be pulled together by a common temporal thread. Daesin is that for which the question of the meaning of being is a question. The basis of Daesin is care, and the structure of care is temporality. In other words, Daesin is that aspect of being that has the question of the meaning of being as an inherent part of it, and the basis of this aspect of being is care, which is structured in time. Daesin is temporal subjectivity for which the question of the meaning of being is a question. Daesin's temporality is the primary phenomenon, relative to which other concepts (including "world time", the publicly communicable, objective time concept used, for example, in physics) are derivative. Time is continuous in a way that space is not. Heidegger argues that even to conceive of time as an infinitely divisible set of instants, a never ending collection of "nows" (i.e. to think of time as continuous and infinitely divisible, as, for instance, a physicist might), our consciousness is already "free-floating" in a temporality, which that kind of concept of time depends upon. In other words, subjective temporality is a prerequisite for the kind of objective infinitely divisible time concept used in sciences and the practicalities of everyday life. The depth of subjective time is a grounding from which we may derive "objective" (scientific) time. It does not appear that Heidegger rejected the idea that space could also be conceived as infinitely stretching in different directions, as he discusses of time, but rather that he regards the continuity of time as more significant than that of space, and closer to the core of human subjectivity. For instance, in tool use, that which distinguishes the ready to hand from the present at hand, "letting a thing be involved" in a task or activity necessarily temporalizes that thing (adds a "time-ness" to that thing, causes it to engage with temporality). The temporality of Daesin arises because Daesin is fundamentally something that stretches out between its birth and its death, and this "between-ness", this stretching out, is the structure of care. Even though different people can agree on a convention by which "public" objective time is measured, and share this consensus about "what time it is", for Heidegger the key point is that at heart these people would assign to this moment in time a different relative significance (dictated by their respective births/deaths, the events of their lives with particular significance, etc). It is this subjective significance that forms the more foundational, real time (relative to which the kind of objective, consensual, public time, used in things like science and measurement, is derivative). Datability of things relative to one's own experiences is, simplistically speaking, "more real" than some relation to the movement of a clock - because that subjectivity makes us what we are. Our basic understanding of time is a "time reckoning" by which we relate events to our subjectivity, it is from this that derivative, objective, public time is derived and developed, rather than the other way around. Public, datable, objective time as is used in measurement, gets leveled off, homogenized. In public time, each "now" lacks "datability" relative to events in Daesin's own subjectivity, and therefore lacks "significance" relative to that subjectivity. Public time is vulgar, ordinary, derivative, and covers up real time. When it comes to the difference between authenticity and inauthenticity, again the nub of the difference between these modes of being is temporal. Anticipatory resoluteness is a being towards death, an anxious but courageous confronting of one's own mortality, which goes hand in hand with a remembering of the temporality at the core of Daesin. Inauthentic modes of Being are lost in the thrownness with which Daesin is in the world, reacts to underlying anxiety with proximal fears, is distracted by the present and washed along in the they-self. Inauthentic Daesin has lost its genuine temporality, and is trapped and confused by a sequence of "nows" without a coherent thread running through them. If authenticity is a way of being what we really are, and inauthenticity is a losing of that primordial authenticity, then the difference lies in how the modes of being interact with time. In short, inauthenticity is lost in the present, whereas authenticity truthfully connects birth and death in an individual, and inherently temporal, manner. The public they-self hides from death because it is individuals and not the they-self, that die. The closest Heidegger gets to (explicitly) touching on a concept like a spirit or soul is in relation to a critique of Hegel's concept of spirit; which again Heidegger proposes as a fundamentally temporal thing. Heidegger famously leaves Being and Time incomplete, saying in essence that he had successfully shown temporality to be the basis of human being in the world, but leaving open the possibility that temporality grounded the question of the meaning of Being in general, i.e. over and above the specific considerations relevant to Daesin (which is necessarily the grounding of any philosophical inquiry, but is not the totality of Being, or its meaning). He never addressed the question of this totality, but he did say that time was the horizon in which an understanding of the question was disclosed to Daesin, and asked the question as to whether time is the horizon of the meaning of being itself.

> *"Temporality has manifested itself............as the primordial structure of Daesin's totality of Being."* *BT*436
> *"Does time itself manifest itself as the horizon of Being?"* *BT*437

After wading through Heidegger's complex vocabulary, and indeed even after accepting his assertions about authentic individuality and the dangers of being washed away in a public collective madness, it still seems that we are left with more questions than answers. Beyond Heidegger's immediate emphasis on interactivity, individuality, and authenticity, the theme that underlies much of his advice is one of recovering a basic relationship between our individual subjectivity and time, a basic relationship that seems to get lost in everyday human life. The difference between authentic and inauthentic modes of being seems to be that authentic modes of being lead to a successful recovery of this underlying temporality; an honest relationship with our own mortality and inevitable death, and individualization of our historicality, and so on. But why is time so special for our subjectivity? Why is truth unveiled/disclosed in time, not just consistently there? What actually happens in our consciousness in a moment of vision? Where did we get this pre-conceptual awareness of the meaning of Being from? Why does interactive tool-using deserve an entire different category of Being that differentiates it from inanimate, non-useful, objects? And most importantly, what is it about us that gives us the fundamental temporality to which Heidegger repeatedly refers?

Previous comparison between Wittgenstein and Heidegger: "*Groundless Grounds*" by Lee Braver

Braver argues in "Groundless Grounds"[32] (*GG* below, italicized text is direct quotation), one of the few explicit, detailed comparisons between Wittgenstein and Heidegger, that the two thinkers share an emphasis on an "original finitude", an underlying whole-ness for which no foundation is possible, but which necessarily guides any system of thought. Braver proposes that the following relatively small number of related, overlapping concepts provide important areas of agreement between the two men's respective works:

Implicit understanding derived from interaction. Understanding derived from the reality of normal interactions clears up the artificial confusions that result from contemplation separated from the real world. Heidegger distinguishes inconspicuous present-at-hand things from the purposeful ready-to-hand objects with which interactivity is possible, and sees it as problematic that traditional metaphysics has been predominantly confined to the former. Authentic modes of being are ways of interacting with the world that reflect a "mastery" of our true nature and its connections to what is outside that nature. For (later) Wittgenstein, when we forget about attempts to produce language with some kind of transcendental foundation, and just use words to interact with others, we understand what words really are - and the pseudoproblems that result from attempting to enamour words with such perfection simply evaporate away. Once we have a context, naturally provided by the kind of situation in which we necessarily live, and a methodology appropriate to that context, we are already a long way to having all the understanding we need.

> "*Heidegger's early and Wittgenstein's later conception of philosophy both use the understanding tacitly manifest in our mundane interactions to correct the misunderstandings derived from disengaged contemplation. This understanding- primarily of word usage for Wittgenstein, and modes of Being for Heidegger- is conceived of as a mastery, knowing how to speak or act appropriately. Constantly in use, it remains inconspicuously in the background and vanishes under the theoretical gaze. False problems occur because the understanding that could prevent them resists thematic examination and articulation.*"GG52

Holism, teleology and tautology. If we have an implicit understanding from our natural activities it is because such activities are purposeful. Purpose is defined tautologically relative to the systems by which we get to it, and there is no need for anything different. Braver notes how both Wittgenstein and Heidegger create an alternative to the confusion that arises from philosophy's attempts at an excessive separation between contemplation and the world, by creating an irreducible holism; an unapologetic circularity, within their own systems of thought. For instance, Daesin is a category of being, authentic or inauthentic in "mode", but authenticity and inauthenticity, (and modes of being as an idea), are by definition characteristic of Daesin, and occur by virtue of it. Heidegger asserts things as being true by a circular definition, rather than defining things on the basis of starting foundations that he takes for granted. Similarly, Wittgenstein's circular, logically consistent system of meaning objects built on names exhibits just as much necessary, unproblematic circularity. Tautology is necessary for meaningfulness. And holistic teleology, "just is" as close as we are going to get an accurate description of real experience, by virtue of its fundamental groundlessness (see below).

> "*Rejecting Russell's conception of logic as comprising maximally general laws of reality, Wittgenstein views it as a set of empty tautologies which, saying nothing, show the structure of and relationships among the pieces of propositions that do say something.*" GG57
> "*The understanding of being that Heidegger incessantly worries about ......is the ability to interact appropriately with the various types of entities that we're constantly interacting with.... It is because Daesin possesses this understanding of being that Heidegger focuses his inquiry into being on this one being.*" GG75

Knowing by living. Braver suggests a parallel between Wittgenstein's forms of life and Heidegger's Being-in-the-world. We learn language through practice and socialization, and we interact with the world in a contextual manner that is simply an extension of what we are as beings. Allowing ourselves to simply live in the world supersedes any understanding derived from holding an esoteric concept up on a pedestal. Just as Wittgenstein attempts to disabuse us of the "reification" of words away from their natural context, so Heidegger argues how knowing is based on Being in the world. As above, linguistic rules cannot be justified without appeals to other rules, and ultimately an unsatisfactory infinite regress. Authenticity as a category into which modes of Being can be divided, cannot be defended on *a priori* logical grounds - Heidegger just asserts, subjectively, that when we adopt and authentic mode of being in the world we recover aspects of our being that we have a tendency to lose in the messiness of human interaction. There is no need to appeal to detached contemplation in order to justify why this is the case, and it would be no use even if we did. In other words, one way or another, both Wittgenstein and Heidegger tell us to avoid getting so hung up on words or detached ideas that we forget to live.

> *"(1) The fact that intelligent speech and action take place in the absence of vigilant consciousness demonstrates that it is not needed, dispelling our compulsion to retroactively implant it. (2) Any reasons I could cite to justify the way I follow a rule would be equally open to various interpretations, triggering an infinite regress of interpretations. (3) Making understanding an entirely interior affair has absurd consequences which conflict with our normal use of the idea of understanding."GG*139

Groundlessness. Braver's central argument is that both Wittgenstein and Heidegger point out how the frame of reference for our thought and our ideas is something that we "receive" as an innate part of us, and as such is not something that we can control or justify by an appeal to something outside that thought. I.e. our thinking can never be "grounded" once at for all, because that ground would itself require contextualization and justification. Foundationalism, the attempt to trace all knowledge back to a single source, is exactly what Wittgenstein rejects when he changes his mind about the *Tractatus'* arguments concerning logic reflecting the fundamental structure of the world. And foundationalism's absurdity is exactly where Heidegger starts when he emphasizes the fundamental interactivity of Daesin. The concepts used to provide a grounding for secondary, derivative ideas, are themselves without grounding. There is no reason why we are Daesin, why we are thrown into the world, why the basis of Daesin is care, why the structure of care is temporality, why the they-self or moments of vision exist, or why truth is disclosed to us temporally. That is just the way it is (as Heidegger defines it). Similarly, Wittgenstein makes no attempt to explain or defend why meaning is use, why language games occur and evolve, why the linguistic aptitude is innate to human beings, why forms of life gravitate towards meaningfulness over time, and so on. Both men would presumably be dismissive of a skeptic who attempted to apply a foundationalist mindset to their ideas as having fundamentally missed the point. For both Wittgenstein and Heidegger, a small number of central concepts provide a ground. This ground is solid, and can be built on in order to take one's thought in different directions. But it is futile and meaningless to attempt to provide a grounding for the initial ground itself.

> "*Wittgenstein and Heidegger challenge a number of the assumptions and aspirations that have guided philosophy since its inception. One of these, foundationalism, is the attempt to trace all knowledge back to a source or set of claims that, as necessarily true, secure the truth of all that is derived from them.......*" GG173
>
> "*Heidegger and Wittgenstein seek to change the way we think about justification. Both argue that we receive our ways of thinking and cannot control them. Both employ infinite-regress arguments - attempts at justifying our ideas either go on forever or come to a halt in something which is itself not justified- and circular reasoning- attempts to justify our basic ways of thinking must use these very ways of thinking, and so can provide no independent legitimation. This encourages a kind of humility....Both philosophers emphasize our dependence on society, nature, and our own human nature - a dependence we never outgrow.*" GG221

Original finitude. Braver's synthesis centers around the suggestion that Wittgenstein and Heidegger both build their philosophies on foundational, irreducible limits. Wittgenstein continually wrestled with the possibility of knowledge of the limits of meaningful language (and thereby the limits of expressible knowledge itself), and the only way out he could find was via the inseparability between forms of life and language games. Heidegger begins with an acceptance of this inseparability from the world. Braver argues that both men followed Kant in accepting the finiteness of our existence and embracing it as important - rather than attempting to explain it on the basis of something more "ultimate." Original finitude can be thought of as the necessary discreteness of groundless grounds. The crucial example of original finitude in Wittgenstein is the "unsayable"; the notion that there is a constraint on what can be expressed in language that we simply have to accept, and pass over in silence. There is no reason as to why this passing over must be the case, the restriction itself is what is foundational. There is no possibility of a successful foundational metaphysics, we must simply derive phenomenological knowledge of how our being in the world is, with all its temporality and messiness. Crucially, Braver notes how the core of both philosophies relies upon a position of humility, on our accepting the unassailable discreteness of the frame of reference in which we find ourselves. Both men recognize and agree with Kant's assertion that this limitedness is a key part of our essence that should be celebrated, rather than something we should attempt to escape, and both dismiss grandiose philosophies that attempt to transcend the reality of our experience. Our grounding in the world is self-evidently limited, finite. This finitude is the closest that we will get to a foundation, and a search for such a foundation is in any case a misguided and deeply unnatural venture - both Wittgenstein and Heidegger suggest that we should just get on with living - something which, though finite, is inherently temporal and interactive.

"Heidegger begins with the sense of finitude that Wittgenstein fought so hard to achieve......'finitude is not some property that is merely attached to us, but is our fundamental way of being. If we wish to become what we are, we cannot abandon this finitude or deceive ourselves about it, but must safeguard it.' This safeguarding informs all of Heidegger's though, beginning with metaphysics." *GG*229

".... Wittgenstein and Heidegger want to create a positive notion of finitude........ They want to examine our limitedness without thereby transgressing our limitations, without peeking over to see that nothing lies beyond the world's horizon- not even nothing." *GG*231

Reflections and suggestions

It is easy to identify differences between Wittgenstein and Heidegger. The obvious place to start is to note that the two men differ fundamentally in the extent to which they are troubled by meaningless language. Wittgenstein never seems to have been truly confident of the meaningfulness of any statement, preferring to relativize different types of meaning to various different language games; different meaning criteria and contexts in which language is used, for a different purpose in each case. By contrast, Heidegger casually invented entirely new words, and seems to have regarded multiplicity of meaning, and by implication linguistic ambiguity itself, as an indispensable part of meaningful language. Heidegger framed this in terms of "the life of language", stating that " *The life of actual language consists in multiplicity of meaning. To relegate the animated, vigorous word to the immobility of a univocal, mechanically programmed sequence of signs would mean the death of language and the petrification and devastation of Daesin*"[33]. This difference in attitude must be confronted and made precise before a comparison between the respective works of Wittgenstein and Heidegger can be attempted or applied to metaphysical questions. One way to pin down Wittgenstein and Heidegger's different attitudes is with the following statement:

"If a question can be asked, it is either answerable (at least in some hypothetical/imaginable situation), or it is meaningless."

Is this statement true? Martin Heidegger would, one imagines, dispute the truth of the statement. For Heidegger, posing unanswerable, even borderline meaningless questions, allows us to "point at" issues of which that we have some sort of intuitive, "pre-conceptual", pre-linguistic understanding. This understanding, by definition, is in some way "prior" to our capacity to express ourselves in language. We have a kind of "feeling" that there is something "underneath" our inclination to ask questions like "What is the meaning of Being?" But if we analyze the linguistic structure of such questions rigorously, according to some pre-existing grammatical or semantic criteria, we are forced to conclude that, strictly speaking, such questions don't really mean anything. The words we use to express the inquisitive inclination that we possess only really scratch the surface. The infinite regress that will result from strict linguistic dissection of the question will inevitably lead to a collapse into nonsense. The issue is therefore whether there is anything worth thinking about left over, after the meaningless language has been swept away? Is there anything to the notion of some underlying "intuitive" or "pre-conceptual" understanding, outside strict linguistic expressibility, but hinted at, albeit inadequately, by words? Is this sort of thing a legitimate idea, or just vague, irrational rubbish? The answer depends on one's opinion as to the truth/falsehood of the above statement.

The Ludwig Wittgenstein who wrote *Tractatus Logico Philosophicus* would certainly say that the statement above was true, and would presumably have also been quite scathing about questions that were only theoretically or hypothetically answerable. In the spirit of the *Tractatus*, if a question is not answerable, and answerable clearly, we must simply accept that it is meaningless, and pass it over in silence. The later Wittgenstein would probably have stuck to this, but would have been less dogmatic than his earlier self, contextualizing the notion of "meaningfulness" to a multiplicity of different language games, developed for a multitude of different situations, each with their own, relative, concept of meaning. Although relativized and evolving, meaningful language is still, for the later Wittgenstein, the core concept that holds legitimate discussion together. One thing that remains consistent throughout Wittgenstein's thought is the idea that the proper role of philosophy is to clarify language so as to maximize meaningfulness. The reason that the later Wittgenstein comes across as less puritanical about meaningless language relates, amusingly, to fundamental problems associated with the "meaning" of the definition of meaningfulness itself. Using the terminology of the *Tractatus*, if names refer to objects, what is an object? If states of affairs are complex configurations of simple objects, what is actually meant by simple and complex? If some kind of "general propositional form" makes itself apparent (as the nature of some kind of universal underlying similarity between meaningful language, and the relationships within the world that the language reflects and describes), how do we differentiate between the general propositional form that tells us something about the logic of our language and the local idiosyncratic syntax that we have to see through? What do we even mean by "formal" and idiosyncratic properties? And so forth. To spend all one's time eliminating meaningless nonsense, one must be very confident of one's definition of meaningfulness. But no definition can ever warrant such confidence- unless perhaps, it is local, relative, and restricted.

Language's natural temporal evolution is another factor that makes the meaningful/meaningless distinction more subtle. If meaning criteria change over time, how does the language referred to by these criteria remain meaningful as this change occurs - and how do we know? Presumably the later Wittgenstein would add clauses to the above statement, which introduced a relativity and fluidity to the concept of meaning. Any more universal concept of meaningfulness must ultimately be defined by virtue of a "groundless"[25] criterion which is itself not justified or defensible, but merely acknowledged as an unavoidable assumption. If one takes the above highlighted statement as true, what does one use as a definition of meaningful? How exactly does one decide that there legitimately does exist a situation in which a question is (potentially or actually) answerable, and what criteria does one use to make this judgment?

Linguistic clarification works by relating words and symbols to the use that grounds them in the world, and by sweeping away combinations of symbols and words that move away from that grounding, i.e. that take the language too far away from the world, and thereby produce meaningless pseudoproblems. Wittgenstein's contribution to human thought was to demonstrate this basic requirement for meaningfulness, as the homology between language and the "world-facts" being spoken about by language. This basic idea applies to questions like "Is the good more or less identical than the beautiful?", or "What is the meaning of existence?", to those concerning whether there is any need for mathematical ideas like Godel's incompleteness theorems or Cantor's continuum hypothesis. There are some linguistically meaningless questions to which it is simply absurd to expect an answer. Sweeping them away allows us to appreciate the importance of what must be passed over in silence.

But not all language is meaningless, some words are used consistently and appropriately, within a context that gives them meaning. The natural implication of the distinction is that it should be theoretically possible to sort the matter out once and for all. Philosophy should be able to sweep away pseudo-problems, ending up with a logically perfect language that mirrors the structure of the world, then stop. With appropriate philosophical guidance in problem formation, meaningless pseudo-problems should, eventually, cease to even arise. If meaning is "merely" use, why can't we simply sort out a final consensus about which words are useful, then stick to it? If the purpose of philosophy is to "cure" the philosophical impulse by linguistic clarification, then what is the problem with aiming at a final, fixed, logically consistent, useful, meaningful language? This kind of objective was, of course, one that Wittgenstein came to see as futile (much to the disappointment of parts of the analytic school). He rejected the idea of a universal logico-mathematical calculus that objectively reflects the structure of the world, in favour of multiple, contextual, language games. He envisaged a multiplicity whereby different words are useful in different situations, and appears to have regarded it as deeply misguided to expect anything else. He reached the conclusion that within the language world relationship there was only multiplicity[35] - and that the attempt to impose singularity was deeply un-natural and flawed. Why?

Even if meaning is "just" use, use could still, hypothetically, be consistent rather than evolving. Incorporating Braver's arguments, use forms the "grounds" of meaning, and this ground is still solid, and effective as a foundation, even though it is itself groundless. So why is the ground not permanently fixed? To stretch the metaphor (perhaps a little too far) what causes the "landslides" in meaning as use? Or, to use an alternative metaphor, why is language more like a tool that we use, than a picture that we take? Is this connected to the fact that tools are used over a time interval, for a particular purpose? I.e. to the fact that the use of a tool is a temporal concept, in a way that the representation of an object in a pictorial form is not, because a picture is merely a snapshot in time? But even then, the idea that words are more like tools than pictures doesn't necessitate that the available set of tools at any particular time is unavoidably incomplete: it is quite straightforward to imagine a non-evolving toolkit. Why do language games need to evolve at all, why is this a necessary part of the human form of life? Such considerations boil down to the first key question, which, I argue, informatively links Wittgenstein to Heidegger:

"1) Which property of the world makes a final, universal, temporally fixed meaningful language impossible?"

Heidegger's work engages from the very beginning with the subjectivity that Wittgenstein initially attempted to pass over in silence. As such, *Being and Time* would surely be a definitive example of the kind of meaningless nonsense that analytic philosophy sought to sweep away in the spirit of the Tractatus. But concepts like "they-self", "Daesin", "Authenticity", "moment of vision" and "anticipatory resoluteness" might be considered useful, in the sense of providing some sort of helpful perspective to normal people attempting to confront their own mortality and deal with their subjectivity. Importantly, this is still true, i.e. these ideas could still be useful, even if there is a blatant circularity in how all these concepts are defined relative to each other. Every point at which Heidegger (tautologically) defines something unique or definitive about human being in the world, there is some sort of relationship between this definition, i.e. between the prioritized concepts in Heidegger's terminology, and the fact that Daesin is in some sense "embodied time", that we are temporal entities by our very nature. But, why is this the case? Why are we embodied time, rather than (say) embodied space, matter or "spirit"? Heidegger gives a proximate answer to the question "Why not (say) Being and Space?" along the lines of "Because the basis of Daesin is care, and the meaning of care is temporality." But this is ultimately nothing but a circular restatement of the question itself, which can be instantly rephrased as "Why is this structure of the basis of Daesin temporal?" Similarly, one might ask "Why does Daesin exhibit historicality?" Abandoning the Heideggerian terminology, this issue amounts to the second key question for the comparison with Wittgenstein:

"2) Why is time such a special part of our subjectivity?"

I argue that questions 1 and 2 must necessarily be asked and answered simultaneously. The feature of the world that makes a temporally consistent, logically perfect, meaningful/useful language impossible, is the same feature of the world through which we experience time, subjectively, to the most striking extent. The answer to both the above questions is, I suggest:

"Because novelty is real, and novelty connects meaningful language with time, subjectivity and truth."

Novelty refers to the discontinuity that exists in the distribution of cause-effect relationships within time. A particular novelty is the operation, at a specific point in time, of a causal rule linking two specific changes (where a change is anything with both identity and temporality) - which has never operated at any previous point in time, and which continues to operate repetitively after this first instance of its operation. Thus, an instance of novelty is when a repetitive causal rule comes into existence. Causal rules can link changes happening in physical systems, or they can link ideas in the mind. We can therefore look at the implications of the existence of novelty from both an outward-looking perspective (i.e. novelty in terms of the causation that happens in the world) and an inward looking perspective (novelty in terms of the connections between ideas within our internal subjective consciousness). Both these types of novelty, internal and external, share an underlying directionality - particular changes become consistently associated with each other when previously they were not, and this association between changes is repetitive, is consistent in time, not chaotic (inconsistent in time). Meaningful language necessarily refers to the features of the world that are repetitive, that can be meaningfully classified in terms of cause and effect, (as indeed does any coherent conception of "truth"). If there is anything about time that distinguishes it from the other core features of our experience, it is directionality. It is novelty that imparts directionality to time, and it is the directionality of time that we pick up on in our own subjectivity. In this sense then, novelty provides a kind of "thread", an underlying theme running through the concepts of meaningful language, truth, and the role of time within our subjectivity. This theme will be expanded upon in the various speculative suggestions that make up the remainder of this chapter.

Novelty necessitates the evolution of meaningful language over time

Suggestion (1) ***Novelty provides an underlying explanation for the fact that meaningful language needs to evolve over time, and that an internally consistent, meaningful/useful language can never be permanently temporally fixed.*** *Tractatus Logico Philosophicus* refers to a basic similarity between the internal structure of particular instances of meaningful language, and the internal structure of the aspects of the world to which these instances of language refer. This basic similarity underlies some sort of "general propositional form", which is reflected in all language-world intrinsic homology. A state of affairs is a particular combination of, and collection of relationships between, simple, indivisible, unjustifiable "objects." Objects are that which form the most basic structure of the world, and objects have the possibility of the states of affairs within which they may occur written into them from the start. For a proposition to describe a state of affairs, the relationships between the objects within that state of affairs must be in some way "mirrored by" or "homologous to" the relationships between the constituents of the proposition. Propositions that successfully mirror states of affairs in this way share some sort of underlying structure - a linguistic structure that reflects the structure of reality itself. The totality of all such meaningful language-world homology exhibits a uniformity, which is along the lines of what Wittgenstein means by "general propositional form", a structure for meaning.

This nature of this structure, and the reason for its existence, cannot be pulled apart in language, because language-world homology is something that language necessarily cannot step outside. The clearest example of this idea is Wittgenstein's picture metaphor: a picture can depict a non-physical or otherwise untrue situation, but not a non-geometrical one: because a "non-geometrical picture" is a meaningless contradiction in terms. (As mentioned above, even if mathematicians claim to be able to imagine multi-dimensional systems of mutual constraints, they "imagine" such connections, as opposed to literally visualizing or "picturing" them). Non geometrical, non "picturable" situations do not have, within themselves, sufficient intrinsic similarity to how things may be pictured (in the mind, or on paper etc), for us to in any way describe such situations in a picture (true or false, useful/meaningful or not). Furthermore, one cannot ask questions about (say) why some imaginable situations are representable geometrically and some are not, by means of a picture.

This principle applies universally, not just to the tautological constraint that meaningful pictures must be geometrical. There are aspects of language-world homology or "describability" that cannot be discussed in language - because the reasons for the existence of this homology are "off the scale" of linguistic expressibility, just as a multi-dimensional "hyperspace" is off the scale of a Cartesian xyz axes. This restriction on language is real, is something that we encounter, and something that is of profound importance. But not something that we can talk about. Describing why a general propositional form for language-world homology exists would necessitate stepping outside this form whilst still using language, which will only ever result in meaninglessness. If we adhere to the lessons of the *Tractatus*, we must pass over in silence any inclination to attempt such descriptions, drawing a limit to our linguistic expression, and our thought. This is what Wittgenstein meant by the most important part of the book being the one he left out.

The poetic appeal of this perspective disguises the fact that it is by no means a trivial task to distinguish between the situations in which language is becoming ambiguous but can, with work, be cleaned up, from the situations in which silence is the only meaningful option. It may be fairly obvious when the stringing together of random words produces what is clearly gibberish. But such situations make up only a small fraction of the instances of linguistic ambiguity that some might label as nonsense. For instance, is there really nothing behind questions like "What is the meaning of existence?"? How do we know? What if we defined "existence" as (say) the totality of matter and space, and "meaning" as (say) subjective emotional value? One approach is to be puritanical and dismissive of all such linguistic ambiguity - pointing out (say) the potential for an infinite regress in trying to find meaningful definitions of words like value, matter, space, etc. Within this sort of approach "What is the meaning of existence?" cannot be retrieved and must simply be passed over in silence. But how can we tell, more generally, when we are inadvertently trying to "say the unsayable"? Heidegger's perspective, that some kind of foundational, preconceptual indivisibility is both as real and profoundly significant as it is linguistically ambiguous, is totally irreconcilable with a Tractatus-inspired dismissal of anything that starts to look nonsensical. Linguistic vagueness and ambiguity often reflects, for Heidegger, some sort of underlying metaphysical importance that we understand prior to our capacity to express it in language. It is precisely when language starts to become vague and confused, when we start to press against the limits of language, that things become metaphysically interesting.

Thus, either Heidegger must be dismissed outright nonsense, or there must be some sort of problem with the *Tractatus*' notion of a final, general propositional form underlying the totality of all instances of language-world homology, beyond which silence is the only option. Although it is easy to become impatient with Heidegger's style, here I take the second view. The foundational, ineffable subjectivity at the core of Heidegger's perspective is wrapped up in the same considerations that caused Wittgenstein to change his mind.

The property of the *Tractatus*' conception of language that came to be rejected by the later Wittgenstein was its generality. There is some validity to the broad idea of a propositional form, at least in terms of some intrinsic structural similarity between the proposition and the world-fact it describes. (For example instance the tie between self referential mathematical operations and the fractal patterns such mathematics may describe, or the use of the infinite continuity within irrational numbers like e or π to describe different kinds of continuity). But there is less validity to the concept of a general (i.e. universal) propositional form. Throughout his later work, Wittgenstein talks of confusions arising from attempts to use words that are outside the scope of a particular, local language game. Similarity between language and the world is not the problem. The problem is the attempt to encompass the whole world within a single instance of homology with language. Wittgenstein's later writings instead convey cynicism towards finality and universality. He exhibits fierce skepticism towards transcendental "puffed up" instances of assumed linguistic meaning, which act as "foundations" for meaning as use without actually being a part of useful language. Wittgenstein's cynicism is most obvious when directed towards the ungrounded generality of Godel and Cantor: Wittgenstein simply doesn't accept the need for mathematics to have foundations, because foundations of mathematics are not useful/meaningful within mathematics. The "foundations of mathematics" constitute an example of an attempt to justify a language's utility by using the language's own conventions - to step outside the frame of reference within which that language is appropriate. As such, the notion of a "foundation of mathematics" is comparable with a non-geometrical picture. One can't say "P states that P is unprovable in T" and then go on to say "Suppose P true..." in exactly the same way in which one can't (for instance) draw a diagram of an emotion: there is a fundamental mismatch between the idea being represented and the tools doing the representing. Wittgenstein doesn't seem to accept the need for justification from the outside, for mathematics or any other linguistic formalism. There is only multiplicity in the language-world relationship, and it is artificial and misguided to pretend that there is universality. Within any one language game, to ask how and why that game exists, how and why particular statements mean particular things, is outside the scope of the game - and as such, a meaningless question. To refer to one of Wittgenstein's own more famous examples, when laying bricks, there is no asking "why" the name "brick" corresponds to a rectangle shaped ceramic object, and "slab" to a flatter, square shaped object. That's just how the "game" of basic communication within the task of bricklaying is "played"; making inquiries about the etymology of labels for building materials would be a pointless thing to do whilst working on a building site. Asking for a single, logically perfect language by appealing to some external reference point is analogous to asking why a brick is called a brick, and expecting the answer to be somehow relevant to bricklaying. But why? If we simply embrace linguistic multiplicity how we are supposed to distinguish between the situations in which we are inadvertently stepping outside the current language game, but might quite legitimately want to begin playing a new language game, from the situations in which we are simply speaking nonsense, and silence is the only meaningful response? Can we always keep pushing the boundaries and changing the rules of our language games? Is silence never inevitable? Is there enough multiplicity in the world to talk about anything?

If we can potentially talk about anything, then we can talk about how we talk about different things. That is, we can talk about how different linguistic/semantic criteria, taken from different contexts, are connected. But if this is the case we should eventually be able to completely connect all language games, developing a "language game" for relating different language games. Obviously, such an interconnective language game would be dangerously close to the universal general propositional form which Wittgenstein initially proposed, but later sought to move away from. If no such general propositional form is possible, how do we know when we are misguided in trying to relate different language games? In short, if there is only linguistic multiplicity, is there such a thing as "ultimate, permanent nonsense"? Are there situations that are not amenable to description in any language game, and if so, how do we tell what that are? Does there exist the category of "that whereof we cannot speak" at all, or can anything be potentially described in meaningful language? If there does a permanently ineffable category, does the boundary of objective nonsense not delineate a general, meaningful language-world structure that unifies all the multiplicity of the world after all? If we decide that we cannot talk about everything; there are some situations for which we cannot develop language games. What are they? If something is nonsense in the local context of the language game being played, does this automatically make it nonsense outright? If so, every time we move outside the boundaries of a local language game, we are compelled to pass the matter over in silence. Except the silence in question does not profoundly reflect the structure of our world like the silence of the *Tractatus*- it is just "not the done thing" within the game in question. Crucially, where precisely do the boundaries and rules of usefulness within language games come from? Where do language games come from in the first place?

Tacit awareness of the natural temporal evolution of language is a clear feature of Wittgenstein's later writings, and a natural extrapolation from the multiplicity and "family resemblance" properties of language to which he refers. Families resemble each other because of the mutual derivation of family members from a common ancestor, which is an inherently temporal relationship. Language-world relationships do not have one single property in common, just overlapping sets of properties. This naturally gives rise to the related question "Why do languages need to evolve?" A link between new situations and the use of words is well appreciated within the contemporary academic philosophy that has picked up upon Wittgenstein's thought. Stanley Cavell argues that expertise leads to the ability to apply a word in a new situation[34]. Knowing the meaning of a word amounts knowing both the context in which that word can be applied, and any further contexts into which that word can be projected. Cavell interprets Wittgenstein as meaning that one has mastered a word, and the grammatical rules in which it may be used, when one can apply this word to a context that differs from the context in which one initially learned it. He distinguishes this such of natural projection of a word into a context for which it is appropriate, from a more nuanced and technically/literally inappropriate projection, such as a metaphor, referring as follows to the difference between metaphorical and actual transfer of a word to a new context: "*..what is essential to the projection of a word is that it proceeds, or can be made to proceed, naturally; what is essential to a functioning metaphor is that its 'transfer' is unnatural - it breaks up the established, normal direction of projection.*"[36] This is certainly true of how we learn language. But I suggest it does not go far enough: There is no getting away from the fact that we need to invent new words, that the reality of time as we experience it demands that we encounter objects and situations for which there is neither a sound nor a sign nor a symbol, and we need to invent one. This is the reason that Wittgensteinian forms of life are fundamentally temporal.

When one feels the inclination to say something "nonsensical", i.e. outside the language game defined by the local context, one has three choices. First, self-correction and silence. Second, change one's situation to a context in which the rules of the language game are more appropriate to whatever one feels compelled to say. But if neither of these options work, as is likely in some instances, then there is a third choice: invent a new language game. This third option is an essential part of the way in which our mind-language-world relation encompasses and accounts for the novelty that we encounter. Novelty, both in the world and in consciousness, is the reason that language needs to evolve - and hence the reason that a general, final, language-world homology is impossible. Meaningfulness evolves, due to the reality of the new.

The language-world relationship in which our thought takes place is itself a function of time - because new causal relationships enter into the world and require linguistic description, and because new ideas enter our subjectivity and change our perspective. (In the latter case, of subjective novelty, one would need to think in terms of causal rules, linking changes, etc, i.e. adopt the labeling system described in the previous chapter, but with subjective changes in consciousness in mind, rather than objective physical changes in the external world). Novelty is the property of the world that (implicitly) distinguishes Wittgenstein's later philosophy from his earlier work: Novelty necessitates linguistic evolution. Novelty is what makes a general propositional form, in the sense intended by the *Tractatus*, an inherently transient concept - any general propositional form would only ever remain general until the next time the world changed. This fluidity within the relationships that exist between constituents of the world is exactly what makes a final, logically consistent linguistic description of the world impossible. A final, logically consistent general propositional form is possible for a discrete time slice - but impossible over an unlimited time interval, because eventually another novelty will come along. The causal relationships introduced to the world by novelty cannot be predicted from, or described in terms of, the relationships previously present. Grammatical rules and semantic conventions must evolve so as to retain an underlying similarity between them and the external rules governing facts in the world. Thus any language must continually change so as to track, and recover an intrinsic similarity to, the facts of the world that it describes.

The temporal evolution of colloquial language is, of course, an obvious feature of our everyday lives. But a reader skeptical of this emphasis on novelty might refer to the more fixed human languages, most obviously mathematics and logic. Mathematics in particular exhibits a timelessness and an apparent deep connection to the physical world that has lead to it continuously being attributed religious-like status since the time of Pythagoras. The consistency of (say) the value of π between the time of the ancient Greeks and today could reasonably be taken as evidence of a genuine underlying language-world relationship that is temporally fixed and has no need to incorporate any novelty. Similarly, many might argue that the generality of the syllogism seems to reflect a real underlying feature of how we think. We have to assume, extrapolate, and deduce, there is no other way to think - and language, timelessly, might be argued to reflect the fixed nature of this constraint.

Such skepticism towards the universality of the need for linguistic evolution is reasonable. There are indeed timeless features of mathematics that suggest an underlying consistency. Even the fact that mathematics "evolves" through additions to it by mathematicians could still be interpreted as discovery of some underlying Platonic consistency, rather than invention. This "realist-formalist" distinction is, of course, an unresolvable question of perspective and semantics. Here, I simply assert that mathematics, logic, and all other languages, do not have any existence outside the minds of those using them, and demonstrably do evolve over time. For instance, the assumption that probability belongs to the same subject as geometry and arithmetic is a fairly late addition to mathematics as a body of thought, and this addition has happened due to changes in the degree of inherent determinism in the circumstances in which people use mathematics. Like all languages, the properties of mathematics depend almost entirely on utility. The "only" reason that "$1 + 1 = 2$" is a constant tautology is that human consciousness has evolved in a frame of reference in which matter occurs within space in discrete, distinguishable pieces, and counting such pieces of matter was the basis for the invention and application of mathematics. There is nothing inherently transcendental (or even, strictly speaking, fixed) about this contingency. For argument's sake, in some hypothetical, pseudo-spiritual situation in which mind became separate from matter, perhaps $1+1=2$ would lose its practical value (although perhaps this also amounts to saying that mathematics itself would also lose its value in such a situation). More seriously, it is quite clear that if and when the circumstances in which humans need to do mathematics change, the mathematics evolves to incorporate these changes. Some languages may evolve less than others, but all languages evolve, because languages need to describe the new things that happen. The same assertions apply to logic, though again faith in underlying Platonic consistency cannot be refuted.

As Wittgenstein showed, a private language is highly problematic, because there is no reference point by which the user can tell whether (s)he is applying the language's rules correctly, and therefore no real criteria by which to define the meaningfulness/nonsense boundary. In a hypothetical world in which language no longer needed to evolve, the last surviving speaker of a previously public language that had "gone extinct" would possess memories of interactions with others, by which (s)he could determine whether (s)he was applying the language's rules correctly, meaning that (arguably) the problems with a private language need not apply. But if the necessity for the evolution of that originally public, now private language, is introduced to such a situation, the private language argument re-emerges. Suppose this speaker of the otherwise extinct language is applying its rules entirely correctly within his/her internal monologue (using grammatical/semantic rules that were acquired socially and therefore have a relative objectivity, assuming the speaker's memory is reliable). Now suppose that some aspect of the world changes, so that the rules of the language must change aswell, in order for it to remain coherent. Suppose the last surviving speaker thinks (s)he has successfully changed the rules so as to cope with this change. But (s)he now has no way to verify this belief, because there exists no other speaker capable of telling him/her whether the language's rules are being applied correctly. Crucially, this is because all previous speakers, the memory of whom may have provided a reference point, spoke the language prior to the change in its rules that has occurred. Only after this change in the rules did the language become private - with all the problems that this entails.

The key argument therefore, is that one of the reasons that a private language is problematic (in addition to the problems of determination of the correct application of its rules), is that a private language is incapable of evolving over time so as to incorporate novelty. Evolution of meaningfulness involves a change in the world that renders aspects of a language meaningless, two or more speakers intuitively recognizing this change, and modifying the language until they both converge (by agreement with each other) on a meaningful description of the change introduced. Without this social reference point, no such convergence is possible. Thus, the inherent evolution of the rules of any and all languages necessarily causes a blurring of the line between meaningfulness and nonsense. When the relationship between language and the world is considered over a time interval, rather than at a discrete time-instant, it is apparent that nonsense is an essential feature of meaningfulness, because it provides an error margin within which meaningfulness can evolve, so as to be able to incorporate the new. This evolution occurs via stages of messy "semi-meaningfulness", through which the public, social, interactive, and inherently contingent properties of language games become indispensable in converging on the new. A complete description of the structure of the world (as defined as the boundary of what must remain unsayable in language) is a possibility, and perhaps so also is some universal "logical-calculus" connecting descriptions of the world that are inherently contingent upon particular contexts. But only briefly, until the next instance of novelty renders that description incomplete. The world changes, qualitatively, necessitating not just the adaptation of word-labels, but of all linguistic rules. Novelty renders a complete linguistic description of the describable world inherently transient.

Novelty is the reason that we can sometimes think what we cannot say

Suggestion (2) *Novel ideas enter our subjectivity in initially inexpressible form, which is why we can sometimes "think what we cannot say", and what makes our consciousness temporal.* The world changes, and language must change so as to catch up with the world, or lose its utility. This process of evolution within the syntax and grammar of language involves repeated cycles of meaninglessness and "fuzzy" communication, eventually resulting in convergence to new definitions of meaningfulness. Language games are social, therefore this evolutionary process must be socially coherent - people must still be able to communicate with each other during the semi-meaningful "evolutionary" phase. Participants in the language game must realize that the world has changed so as to render it meaningless, and they must also have a sense as to the direction the evolution of their game must take, and implicitly agree with each other on this direction. Importantly, this sense of grasping at a new meaning of words must, by its very nature, be outside the boundaries of strict linguistic expressibility. If language evolves, the users of the language must possess a "sense of direction" as to what words mean, even during periods within which the meaning of these words are undergoing change and therefore at least partly ambiguous.

For example, say two construction workers are given a tool with which neither is familiar, but for which they recognize some use. The process of communicating "pass me the X" will begin with gesturing and physical use of the new tool, and evolve to a point at which "X" is replaced by a name for the new device, and an incorporation of this name into the existing rules of the language game. This evolutionary process would not work if the two workers did not empathize with each other sufficiently to pass each other the tool, use it in their labour, point to it, realize when the other was in need of it, and so forth. This sense of direction relies upon an intuitive empathy between participants in a language game, an understanding that both know what the other means. This intuitive aspect is, by definition, is not initially a communicable part of the language game that these participants share at the start of the process. Crucially therefore, the evolution of language requires that we are able to think what we cannot say. This disconnect between thought and linguistic expressibility bears directly on the nature of novelty: thinking of a new idea, coming to a new realization, reaching a new emotional perspective, etc. When we think something that we cannot initially linguistically express, this inexpressibility frequently derives from the fact that our mind is touching upon something new. It is inherent to the relationship between novelty and our consciousness that novelty enters our subjectivity in a form that cannot initially be expressed. The experience of subjective novelty entering consciousness involves an intuitive feeling that there is a new idea emerging within the connection between our minds and the world. This happens long before we have an "original argument" or "novel theory." Subjective inexpressibility occurs first, and this order is important. It is informative to make an explicit separation between the logical and the subjective elements of human thought, and their relationship to language. Thought contains an analytic, logical component that is synonymous with clear, objective/communicable, well-formed language, aswell as an emotional/subjective element that we might colloquially refer to as "feeling" or "intuition." We might say that "thought" is an umbrella term to encompass both these elements of our consciousness, i.e. the sum of linguistic expressibility and subjective intuition. Consciousness is neither logical expressibility nor subjective inexpressibility, it is the connection between the two:

$$analytic/linguistic \overset{consciousness}{\leftrightarrows} subjective/intuitive \qquad (69)$$

In other words, human consciousness is some sort of tie between mind and "heart", logic and intuitive "feeling" etc. Importantly, any sort of linguistic expressibility necessarily falls on the side of logic, the subjective/emotive aspect of consciousness is beyond language. Some might turn to art to attempt to convey this non-linguistic inexpressible element of the human condition - precisely because art by its very nature is ambiguous. If one can successfully express something in a consistent manner, the source of the expression is not one's subjectivity. If one is genuinely speaking (solely) from one's subjectivity, others can never authentically understand.

Thus, consciousness is the tie that exists between these two irreconcilably different aspects of human being. This tie has two properties that are of fundamental importance; individuality and temporality. Expressibility is an individual property, referring to an individual human consciousness- relationship (1) can only ever hold for what is going on inside a single individual mind. The public, aggregate position of this expressible/inexpressible boundary, as far as it refers to real language that is actually used in populations of individuals, is a kind of bell-curve like average, delineated by what we cannot speak of, or more precisely what we cannot speak of coherently to each other. Temporality is also inherent to the subjectivity/expressibility tie within consciousness: what any given individual can and cannot express changes over time. Relationship (69) refers to an interactive process, by which our thought jumps in and out of language. Thought "does work" upon what is subjective, so as to render aspects of it expressible. Temporality therefore exists in the subjective/expressibility boundary as a direct result of the fact that novel ideas enter our subjectivity and require thinking about in order to make coherent.

Suppose our minds literally were organic computers. This would amount to saying that they repetitively followed certain sets of instructions in ordered sequences, performed certain comparisons and calculations, produced certain outcomes that would dictate thought and behaviour according to hard-wired criteria, relative to which the output of these calculations were assessed. Our minds would be logical flow charts combined with optimality criteria. If our minds were computers, we could still quite easily encounter newness in the world - computers make perfectly acceptable explorers. External novelty (i.e. "world novelty" in terms of cause and effect relating matter within space), could be dealt with be dealt with relatively easily by such computer minds. We could compare anything unfamiliar to that which we had a memory record of, through a hierarchical, logical process through which we could decide what the new features of the world we encountered actually were, relative to pre-existing comparative criteria. This kind of process is, obviously, exactly how computers can be and are used in exploration, such as the use of robots to map unfamiliar physical environments. But, equally obviously, a computer has no linguistically inexpressible aspect, no need for art or some other appeal to the ineffable. The instructions that a robot might use to explore a context featuring some kind of external novelty are just that, instructions - arranged in a logical/hierarchical manner dictating the appropriate response to particular stimuli, much like a flow chart. Even situations in which machines "learn", by modifying the relationship between the instructions followed and the stimuli received, learning cannot occur indefinitely. There is always some pre-set process that defines and sets boundaries on the "learning" process. Machine learning must follow a learning algorithm, dictating how the instruction-stimuli relationship is allowed to "mutate." Yet the fact that computers are perfectly capable of dealing with external world-novelty, implies that such external (matter-space, cause-effect) novelty is not the reason for the existence of a boundary between what is and is not linguistically expressible, nor the explanation for why the position of this boundary evolves over time. This temporality within our subjectivity is something more internal. Human minds contain an analytical/algorithmic aspect by which problems can be solved repetitively and hierarchically. Human minds can do anything computers can do (just not so fast, and with less repetitive efficiency). But human minds are not organic computers, "Artificial Intelligence" has failed. This, I suggest, is directly related to the fact that computers cannot innovate - precisely because innovation entails an encroachment of the subjective onto the rational. It is a definitive feature of psychological subjective novelty that it enters consciousness in an initially inexpressible form. Only after thought "does work" to subjective novelty does it (in some but not all instances) become objectively expressible to others through logical/rational language.

The most well known Heidegger scholar is Hubert Dreyfus, and one issue of central relevance to novelty within psychological subjectivity is Dreyfus' discussion of machine intelligence, AI. Dreyfus argued[37,38], inspired significantly by Heidegger's emphasis on subconscious interactivity, that human thought/consciousness is in some way "in" the world aswell as in the brain - that the human mind has an intuitive sense of situation that cannot be reduced to symbolic representation in the logical languages by which computers work. Dreyfus challenged the foundational assumptions on which the artificial intelligence project was based: that the brain processes information in discrete binary operations, that all knowledge can be formalized in logical language, and that knowledge occurs in discrete, autonomous pieces: independent facts that can be represented within independent symbols. Because these assumptions were poorly thought out, and ultimately both unjustified and unsupported by the reality of the experience of having a human mind, being a human in the world, and because the AI project attempted to replicate human thought by manipulation of binary non-interactive memory storage, this project was destined to fail. Dreyfus challenged the optimism with which AI researchers pronounced that machines would soon be capable of performing innumerable supposedly definitively human tasks (the famous examples being the development of theorems in mathematics or becoming the world champion at chess). This challenge was met with anti-philosophical ridicule amongst neuroscientists and computer scientists. But the failure of AI has proven Dreyfus correct, as has the failure to discovery the memory trace (the discrete physical manifestation in the brain of a discrete specific memory).

By analogy to Heidegger's ready to hand/present at hand distinction, Dreyfus differentiated (respectively) "knowing that" from "knowing how." The former, hierarchical, linguistically representable, logical problem solving, can be achieved by computers. But "knowing how" refers to a more subjective awareness of context, which dictates how humans do things during the normal course of their lives. Large parts of human behaviour are an unconscious flowing along the contours of our experience[36], rather than explicit, logical decision making processes. This intuitive, interactive responsiveness to being in the world cannot be formalized in the logic of a computer program. The supposition that there are limits to the representability of the world (i.e. the extent to which consciousness can be exhaustively captured in symbolic language) can be linked to Heidegger's discussion of the temporality of our subjectivity. One might even speculate that Heideggerian interactivity would be more consistent with recent suggestions by Sheldrake[37] that memories are (to some extent) closer to a resonance pattern that the brain somehow "tunes in to" as well as storing them. It should be noted that from a scientific perspective such considerations are highly speculative and the discussion is ongoing. Perhaps the AI project may yet succeed. But the key point is that its failure up until now is at least conceptually consistent with Heidegger's emphasis on the foundational importance of inexpressible, temporal, subjectivity world-interaction. If novelty enters our reason from our subjectivity, AI will never succeed, because computers, confined to the application of objective, representable instructions, will never be able to genuinely innovate.

There is a parallel linking the separation between expressible and inexpressible thought within consciousness, and the separation between chaotic and repetitive cause effect relationships in the world. In terms of causal relationships within the world, those that connect particular changes in an inconsistent manner over time were labeled "chaotic" causal rules, those that connected particular changes in a consistent manner over time were labeled repetitive. At an instance of novelty in the world, causal rules that were previously chaotic (or non-existent, it is impossible to reliably know the difference) become repetitive. I.e., a particular instance of novelty constitutes a causal rule moving from the chaotic category to the repetitive one. Similarly within consciousness, thoughts move from the inexpressible, subjective category (analogous to "chaos") to the objective, linguistically expressible one (analogous to repetition). If one wishes to think of thoughts as causal rules connecting changes within a individual consciousness, then novelty constitutes the origin of such "conscious causal rules." It is impossible to tell the difference between a causal rule in the world that only operates inconsistently and a causal rule that does not really exist, because in order to reliably determine the existence of a given causal rule, one needs to use an inductive reasoning system (like science) that rests upon that rule being repetitive and therefore non-chaotic. Similarly, if one takes the view that a psychological novelty is only real to any degree of reliability if it can be expressed within language, then one can say that it is impossible to tell the difference between a subjective psychological novelty and one that does not exist. Genuine (i.e. reliably existent) psychological novelty is, from this perspective, the process of thoughts moving from the inexpressible sphere to that which can be spoken of. This question of perspective is, of course, dictated by different individual's confidence in the genuine existence of inexpressible aspects of their own subjectivity, and indeed their confidence in the existence of other minds. Regardless, the the key assertion here refers to the order in which novelty enters our consciousness. The intuitive feeling of a new idea, an epiphany, a change in perspective, etc, comes first, expressibility in language follows later (if at all):

$$novelty \longrightarrow subjectivity \longrightarrow expressibility \longrightarrow meaningfulness \quad (70)$$

Linguistic expressibility occurs after thought has done interactive work on the inexpressible "germ" of the subjective novelty:

$$novelty \longrightarrow subjectivity \quad \overset{consciousness}{\underset{\longleftarrow}{\longrightarrow}} \quad expressibility \quad (71)$$

This order is what causes the boundary between what we can and cannot express in language to change as a function of time, what causes things that previously evaded description to become amenable to the constraints of logical language. The size of the objectively describable sphere of experience increases over time; logical language advances. Unfortunately, there is no way to tell what is happening to the size of the incommunicable sphere of experience, and it seems that adding to this incommunicable sphere is a deep and fundamental aspect of human consciousness, so the temptation to label this advance as progressive should perhaps be avoided.

Novel synthetic a priori knowledge evolves into analytic a priori linguistic convention

Recall the dichotomy: either there are no true philosophical problems, merely the misuse of language, or there is something fundamentally important about ambiguously defined but intuitively understandable questions such as that of "the meaning of Being." This dichotomy can by bypassed by a perspective that acknowledges the reality of the new. For instance, knowledge must be thinkable and (realistically) also "sayable", or concept of knowledge becomes meaningless, a consideration relevant to the idea of synthetic *a priori* knowledge. If knowledge is synthetic it cannot be within the language of the propositions used to refer to it, else it is analytic. This, means that synthetic *a priori* knowledge must start out outside of language, i.e. outside the proposition or question that refers to it. In some cases this may amount to what subsequently becomes synthetic *a priori* knowledge initially starting out as "unsayable." Therefore, synthetic *a priori* knowledge must pass from being initially unsayable (because if it is always sayable it is analytic) to sayable (because if it remains unsayable it is not knowledge).

To take Kant's example, imagine for the first time some ancient man noticing that $7 + 5 = 12$. Regarding this discovery as synthetic *a priori* knowledge amounts to regarding the concept of 12 as becoming "sayable" as a result of this discovery. If literally no person had performed this addition before, "12 -ness" would not be something people could speak about, it would not be a concept that would make sense in communication, the concept of $7 + 5 = 12$ would not be in the "language game" of arithmetic.

But equally, it would be incorrect to regard the discovery that $7 + 5 = 12$ as existing in the world "*a posteriori*" to experience. The synthetic act of performing this addition moved the concept of 12 from a vague subjective hunch (which might have caused the discoverer to do the sum in the first place) to "sayable" knowledge. One could regard the point of psychological/intellectual novelty in which the concept of 12 is invented as an act of the conscious discovery of synthetic *a priori* knowledge. But after this point of novelty, because the language evolves over time so as to incorporate the new discovery, $7 + 5 = 12$ becomes analytic, true by definition within the conventions of the language. The evolution of the conventions of a language means that the instantaneous discovery of synthetic *a priori* knowledge is possible, but after this discovery the conventions of language evolve so as to make the new discovery true by definition. The evolution of language naturally allows what one regards as "true by definition" to be a function of time. The key point is that simply thinking about the fact that time is a factor in knowledge, that people have new ideas and work things out via psychological novelty, helps to show how the question of whether synthetic *a priori* knowledge exists can be considered a contradiction in terms. If one allows the existence of novelty in the "mind to language to knowledge" relationship, the existence of synthetic a priori knowledge amounts to the trivial statement that something must go from being "not knowledge" to "knowledge", which exposes it as a pseudoproblem. In terms of the original problem, one might say that if knowledge is representable in language, it is, by the definition of meaningful language, analytic. But what is (potentially and actually) describable analytically in meaningful language changes as a function of time: which is about as synthetic as thought can get. The synthetic *a priori* knowledge problem does, however, imply that we must be able, somehow, to think what we cannot say.

If we assume that we have knowledge that we cannot express in language, then we must in some way be able to think what we cannot say, because we must be able to think about knowledge if we genuinely have it. This, obviously, exposes the idea of "ineffable knowledge" as a highly confused concept. Alternatively, if we assume that inexpressible knowledge is impossible, then once we genuinely have knowledge it must be expressible in language. If synthetic *a priori* knowledge is possible, we must be able to manipulate propositions made in language, so as to derive knowledge that was not already present in these propositions, or in the rules of the language. If this is the case, then in between the time at which we are manipulating the propositions, and the time at which we have derived linguistically expressible knowledge from this manipulation, the knowledge must somehow come to be expressible. Either way temporal change needs to be built in to our concept of knowledge. If this transition between inexpressibility to expressibility is possible, and given that the knowledge going through this transition is thinkable, it must be possible for us to think something that we cannot express in language. Knowledge (that is both thinkable and sayable) can come into existence by the interaction between the expressible and inexpressible aspects of our subjectivity, and the proposition from which the synthetic *a priori* knowledge can be derived. The reality of novelty means that there is no problem with this idea of knowledge "coming into existence", if we allow ourselves to deem an aspect of our thought inexpressible. With this emphasis on time and novelty, *a priori* knowledge is both synthetic and analytic, during and after this process of coming into (expressible) existence respectively.

The attitude towards time that Wittgenstein developed during the course of his later philosophy and informal writings was discussed by Hintikka in "Wittgenstein on Being and Time"[39]. Wittgenstein apparently made a distinction between measurable, public chronology, which he called "information time", and second category of more subjective, but also more foundational, internal, phenomenological, "memory time." Hintikka notes that these two categories of time have different implications for what we consider to be an object, i.e. something to which we can can attribute properties and quantities. If our basic notion of time is information time, objects must persist physically, and undergo change between different instants - information time objects are physical, and change "in" time. If our basic conception of time is memory time, objects are parts of slices of the world and/or our consciousness, which we experience and have an inherent "time-ness" to them. Objects must be time slices, chopped out of a relationship between memory and the continuous temporal flow of experience. The difference is primarily one of perspective rather than substance. One might crudely simplify "information time" as meaning that the essential stuff of things is physical events in the world, whereas in memory time the essential stuff of things is events within consciousness. Tellingly, the Wittgenstein who wrote the *Tractatus* appears to have been of the view that "*what solipsism of the present moment means is quite correct*", i.e. that only the present experience of one's own mind can be relied upon as being real. This is read as meaning that objects must have a temporal-ness to them intrinsically, as opposed to "being arranged in time", or referenced against time, as in a more physics-based time concept. In this regard, Hintikka notes that:

> "*if memory time were the primary one, then the rock bottom of temporal order would be, not a succession of events in physical time, but a relative ordering of my memories and expectations into earlier and later ones. This is what Wittgenstein means by saying that in memory time 'there is only earlier and later, not past and future'.*"

Most importantly, Hintikka notes how Wittgenstein realized that his initial picture theory of the meaning of propositions necessarily applied only to the present - therefore justifying the notion of memory time. For Hintikka, if a meaningful proposition were tied to "information time" it would need to intrinsically contain the capacity for comparison with other instants of time - whereas in reality, all we really have is one complete picture of our experience, then another (different) picture, and all we can really say is that one is earlier, one later (as opposed to one in the past, one the present):

> "*The only reality that can be at the receiving end of the Tractarian comparisons with language is the reality of the timeless now. Otherwise we would need such temporally extended comparison processes which the completeness of the picture view was seen to exclude. And this idea of the momentary character of language world comparisons obviously is precisely the primacy of memory time. In brief, the completeness of Wittgenstein's so-called picture theory of language presupposes the primacy of memory time. Our language means what it means in virtue of saying something about what is now the case.*"

Wittgenstein's subsequent rejection of the picture theory in favour of public, comparative linguistic meaning, is interpreted by Hintikka as indicating a change in attitude towards the primacy of information time in Wittgenstein's later thought, although it is again worth emphasizing that the two perspectives are "*two methods of identification, not two classes of entities.*" The key tie is between Wittgenstein's change from picture like, fixed, to evolving language, and to a comparable change between perspectival (memory) and publicly referential (information) time. Wittgenstein came to realize that public, not "phenomenological" language was the only meaningful kind, forcing a switch to (publicly communicable) information time, "*the time we live in is memory time, but the only time we can directly speak of in our language in physical time, that is, information time.*" Finally, Hintikka argues that this change in Wittgenstein's attitude to the relationship between time and language implies that, at root, there may simply be something inherently subjective and "perspectival" (and therefore presumably in some sense incommunicable) about time:

> "*The problem of integrating different perspectival worlds into one common public world will then become Einstein's question in his special theory of relativity of whether local time can be combined into one common absolute time.......we know that local times cannot be combined into one absolute time......we cannot even expect in general that there is a unique public system to serve as the reality our language is about. And this in turn suggests that the public system cannot be the primary one epistemologically, even if it may be the basic one in the semantics of our actual language.*"

This divide between the two time concepts is, of course, is extendable beyond language or time, space, or any other communicable frame of reference. The potential for separation is nothing more than a reflection of the fact that we experience our subjectivity most directly, whereas in language we necessarily try to be objective and communicable, thereby tied to reference points through which we can interact with the subjectivity of others, via the world. In language, we unavoidably step outside of the most basic, conscious, inherently private elements of our experience: elements that are doomed to forever remain incommunicable to others. As Hintikka says "*the aspect of reality that our language primarily speaks of must be the physical world. It cannot be a phenomenologically construed world. Therefore a phenomenological language is, as Wittgenstein asserted on October 22nd 1929 'absurd'.*" The relevance to the main arguments on novelty presented here are less subtle, and rely only on the self-evident nature of this subjective-objective, private-public dichotomy in our consciousness. The fact that we unequivocally have both subjectivity and communicable objectivity, and the fact that Wittgenstein became confused and underwent a change of heart as to the boundary between the subjective and the objective, exactly regarding how this boundary relates to language, is important. It is important because it is not clear whether language relates most directly to our subjectivity or to what might be termed the "externally objective." The reason for this lack of clarity is that language must do both. Language must be objectively communicable, but it must also flow with novelty, both in the world, and in our consciousness. And novelty in our consciousness occurs, necessarily at the boundary between the subjective, and the objectively linguistically communicable.

Subjectivity, meaning, use, nonsense and gibberish

Novelty explains the significance of time within our subjectivity, as intuited by Heidegger. and this basic argument permits a minor relabeling of certain ideas relating to Wittgenstein's "meaning is use" theme. If something is expressible we can say it in language. If something is meaningfully expressible, we can think it, say it in language, and use it within a language game through which we objectively and consistently express our thoughts. Meaningfulness is thus the intersection between thought, linguistic expressibility, and use. All meaningfulness is useful; the intersection between meaningfulness and uselessness is empty "$Meaningfulness \cap (Usefulness)^C = \emptyset$" (using the pseudo-formalism $(concept)^C$ to mean the complement, i.e. outside, that which is not contained within that concept, and using $\emptyset$ to mean an empty category), and if something is useful this makes it meaningful:

$$Meaningfulness = Thought \cap Expressibility \cap Usefulness = Usefulness \tag{72}$$

An interesting (although not particularly important) question is whether there is anything we can both think and express linguistically, but not use in a language game?

$$Uselessness = Thought \cap Expressibility \cap (Use)^C = \emptyset? \tag{73}$$

This is a subjective and semantic question that depends on what one takes to mean "useful." If an idea occurs in the overlap between thought and linguistic expressibility, an internally consistent system is possible in language that corresponds to this thought (assuming that internal consistency within the language in question is part of what we take "expressibility" to mean). If an internally consistent linguistic expression of a particular thought is possible, then that thought is communicable. Therefore if we take "communicable to another consciousness" as a sufficient criterion for a thought to qualify as being useful, then there is no such thing as uselessness - the category is empty. Alternatively, if by "useful" we mean "can be used in a real, existing, language game by which we interact with the world", then we can decide to label internally consistent systems as useless, but need to decide what we mean by useful. This choice is an arbitrary one, which probably boils down to a question of taste. For instance, Wittgenstein clearly regarded Cantor's continuum hypothesis (that there is no set with cardinality (number of members) greater than that of the natural numbers but less than that of the real numbers) as useless, - despite its internal consistency. But the infinite divisibility of the number line, combined with the fact that both such sets must be infinite, means that mathematicians concerning themselves with set theory can coherently discuss this idea, therefore that the continuum hypothesis is, at the very least, useful in the context of such discussions, (even if it cannot be regarded as "useful" in the same way as the application of mathematics to engineering is useful). Though arbitrary, this distinction is worth making because it allows us to see that uselessness and nonsense are not exactly the same thing. Nonsense, in contrast to uselessness, is the expression of something in language (nominally in accordance with the rules of that language), whereby that "something" does not correspond to a thought:

$$Nonsense = (Thought)^C \cap Expressibility \tag{74}$$

So in this labeling system, Chomsky's "colourless green ideas sleep furiously"[40] is an example of such nonsense, because although it nominally obeys normal sentence structure, and the words all have meaning in other contexts, this juxtaposition of these concept is nonsensical. It is not possible to think of a colourless green idea sleeping furiously without any sort of meaningful idea of the concepts to which Chomsky's soundbite refers. But still nevertheless, "colourless green ideas sleep furiously" is expressible, it contains what are recognizably words with meanings. By contrast, gibberish corresponds neither to coherent thought nor to linguistic expressibility:

$$Gibberish = (Thought)^C \cap (Expressibility)^C \tag{75}$$

Thus a drunk or madman shouting nonsense that does not even conform to words is placed in a different category to Chomsky deliberately constructing a silly soundbite. The last remaining category in this labeling exercise is the most important, because (I suggest) it is the reason that all coherent linguistic systems are fundamentally incomplete. There is a real category, within our consciousness, that we consistently experience as part of our thought, but which we are unable to express in words. We can label this as our subjectivity, the part of our consciousness that we can think of but not express in language:

$$Subjectivity = Thought \cap (Expressibility)^C \tag{76}$$

Subjectivity includes everything from emotional, intuitive, artistic "feelings" up to and including Wittgenstein's "the mystical" (the awareness of the finite-ness of existence, the "feeling of the world as a limited whole"). Importantly, subjectivity is not nonsense. Subjectivity, when considered as a function of time, can become useful - because, by the arguments above, it is from subjectivity that novelty stems, and from novelty that linguistically communicable knowledge stems. The task of thought is to integrate, when possible, this novelty into our existing logical language, express it in a meaningful way using the analytic conventions to which we already have access. Depending on the details of the novelty and the individual, this may be easier said than done. Sometimes an instance of novelty may remain at the level of an inexpressible subjective/artistic impulse, that to a skeptic "doesn't really mean anything." Sometimes an instance of subjective novelty may become integrated into a person's existing thought processes and produce communicable idea that other individuals can meaningfully comprehend. But either way, subjectivity is qualitatively different from nonsense.

Human consciousness: Novelty by the forcing together of reason and subjectivity

Life is the most novel feature of the natural world. This is a direct result of the fact that life pools together chemical systems that function in homologous ways but are composed of different stuff, that are not "supposed" to be together in terms of the laws of inorganic chemistry. There is no reason for nucleic acid and protein to electrostatically/ionically combine together - in terms of energetic considerations, in terms of the chemistry dictating how either are polymerized, in terms of the internal rules as to how either nucleic acid or protein are built intrinsically. Life is the natural phenomenon by which nucleic acid and protein combine. Template replication, the correspondence in sequence between $DNA \rightarrow RNA \rightarrow Protein$ emerges as a result of the spatial compartmentalization of particular enzymatically promoted chemical reactions in time and space. This complementarity, in large part, explains the qualitative increase in diversity in form and function that has occurred during the history of life. Life draws together physical and chemical systems that "should not" be together in terms of the intrinsic rules by which they are constructed. This is is exactly why life is the most novel natural phenomenon. Template replication derives from the forcing together of nucleic acid and protein. The eukaryotic cell derives from the forcing together of internal bacterial symbionts and a cell with a moving cytoskeleton that can spatially compartmentalize the energy these symbionts produce. Symbiotic interactions produce physiologies that are greater than the sum of their parts. Co-operative transitions in the level of selection lead to multicellularity from cells, social groups from individuals, and so on. The novelty generating ability of life derives from the fact that it forces together separate systems that need not interact.

The novelty generating ability of mind is an iteration on this theme. Mind is the most novel feature of life, human consciousness the most novel instance of mind. In an exactly analogous way, the human mind draws together thoughts that "should not" be together in terms of the relationships between the aspects of the world to which these thoughts refer. This "drawing together" makes no sense in terms of the internal rules by which each of these thoughts work and refer to the outside world. But this "symbiosis" between pre-existing thoughts, which are fundamentally different but have some sort of homology between them, is, I speculate, what gives rise to new thoughts, some of which subsequently become linguistically communicable. There is no reason for subjectivity and logic to be tied together. Thinking in terms of the modern machine analogy for the mind, there is no need for a computer or database (or whatever "information age" metaphor is preferred for the brain) to be tied to the site of emotions. The link between analytical/reason-driven thought and subjective emotions is not something that can be explained in terms of the nature of either reason or emotion. Yet human consciousness is, at its very core, this tie between reason and subjectivity. The reason for this tie's existence is at best an elusive part of our evolutionary past, that may one day be appropriately explained by historical science, and at worst an unknowable by product of a genuinely irreducible mind-matter dualism. The reason that the human mind is not a computer is that the human mind mixes computer-like analytical/logical functions with emotions and subjectivity. And it is this connection between subjective and objective that causes human consciousness to be the most novel instance of mind.

Even a skeptical person who wished to dismiss the appeal to the internal novelty of one's own imagination as unverifiable or wishy-washy thinking, would surely have to acknowledge the propensity for novelty evident within human technological development innovation. Human technology, systems of thought, patterns of behaviour and modifications of the physical environment do something to planet Earth that is qualitatively different from equivalent effects of any other species. It is not currently politically correct to attribute any special status to humanity.The modern popularization of science, particularly when it comes to the relationship between humanity and the rest of nature, asserts that the only rigorous outlook is a negationist one, in which humanity has no special status, and to argue otherwise is pseudo-religious backward thinking. Consciousness and "the self" are dismissed as illusory, and the role of contingency in evolution is instead emphasized, chance by-products of repetitive nucleic acid evolution, etc, etc. This attitude stems from a denial that there is anything unique about mind, or about humans. I argue here that the way to appreciate the absurdity of this denial is simply to recall the experience of thinking of something new, or imagining something impossible in the real world. The novelty of mind in general, and human consciousness as distinct from other manifestation of mind, is self evidence. But any "human-specialness" is merely a difference in degree. Apes can use tools or language, and non-human examples can be found every time a supposedly definitively human trait is presented to the public during the course of the popularization of science. Similarly, non-human animals do possess consciousness, and some degree of innovation, just not as much as humans. It is the degree and ubiquity of the capacity for innovation that separates humanity from the rest of the natural world. Humans consciousness, specifically human imagination (as opposed to reason) is by far the most novel feature of mind. The unique thing about human consciousness is the extent to which initially inexpressible novelty emerges within it over time. Why?

Subjectivity disrupts reason. The influence of emotional inexpressibility on our analytical/reason-driven capacities makes us less efficient as computers, disrupts our problem-solving ability. But problem solving is inherently repetitive. When we learn (say) a new method in mathematics, we learn how to "do" that piece of mathematics, apply that method, play that game. Asking why a particular method (language game) is the way that it is, entails stepping outside of the rules that dictate what is correct and incorrect within that method. But this incorrectness is exactly how we create new games. If we try to do something new with a particular analytical set of rules, we make a mistake according to those starting rules. Therefore we cannot be genuinely and completely analytical, logical, reason-driven, and genuinely innovative, at the same time. Innovation comes from the subjective in the first instance, only subsequently do we try to force the analytical elements of our thought to catch up. A good inventor has a "brainwave" first, only later tries to make the contraption work. A good mathematician has an intuitive sense of coherence first, which subjectively drives him/her to make an internally consistent piece of new mathematics, which he(she) tries later to make consistent with the other areas of mathematics. A good artist has a emotional compulsion that demands expression, putting this expression into (say) grammatically correct words only comes later. And so forth. The key argument then, is that the fact that human consciousness is the most novel feature of experience is directly related to the fact that human consciousness constitutes an inherent tie between analytic/logical repetition, and a subjective/intuitive stepping outside of this repetition. Repetition alone is (by definition) not novel. Subjective, emotional stepping outside of existing boundaries is not coherent or meaningful enough to be recognizable as novelty. Both are necessary. The forcing together of the subjective and the analytical is, I suggest, what creates the human consciousness' capacity for innovation. In short, novelty results from a mental symbiosis between subjective and analytical consciousness.

Both Wittgenstein and Heidegger reacted against precursors of what might be called scientific materialism. The clearest example is the above quotation from *Zettel*, which I have interpreted as referring to what we today call the memory trace, but which at any rate explicitly disputes the premise that brain processes "cause" conscious thoughts in any meaningful way, whatever parallels might exist between neurological patterns and qualitative or subjective psychological states. Heidegger's foundational concept, Daesin, creates an entirely unique category of Being for human-being in the world, reflecting how he sees human uniqueness as something fundamental. It is also obvious that Heidegger views the uniqueness of human life as inextricably tied to its temporality. Wittgenstein and Heidegger express their human-centric attitudes in different ways (explicit discussion of consciousness and emphasis on temporality respectively). But both regard the unique properties of human consciousness (being) as so self-evident as to be not worth bothering with. Wittgenstein correctly predicts the failure of "neuro-psychology" (let alone artificial intelligence) in a casual and dismissive manner in the 1940s. Heidegger famously picked up on on the stresses and sense of lostness that a technological age will bring to modern life[4], well before the internet age. Both men thus appreciated that there was some fundamental aspect of human-ness that inherently becomes lost in reductive expression. But problems also arise with too much human uniqueness, exemplified, of course, by the contradictory and vague nature of any attribution of this uniqueness to a religious/spiritual conception of spirit, and the logical problems that arise from any mind-matter dualism. If mind and matter are fundamentally different kinds of things, how can matter cause changes in mind and/or mind cause changes in matter? If mind is not reducible to matter, why is the presence and correct neurological operation of the brain obviously necessary for the manifestation of mind as a phenomenon? If mind is reducible to matter, how does mind come about through matter? If the answer is "emergence" - what is "the emergence of mind" other than a restatement of the question, and where do phenomena like animal instinct come from? If mind and matter are both reducible to something else, and this something else is neither mind nor matter, what is this substance and how do we know that is exists? None of the answers to any of these questions are known, despite the aggressive and over confident claims of what might be called "atheist-materialism." It seems that both Wittgenstein and Heidegger (as I read them) appreciated that all questions of this sort are self-evidently poorly framed, and unanswerable - but that there is still something "else" that indisputably sets human subjectivity apart. For Heidegger the uniqueness of human beings derives from an intrinsic historicality and temporality. For Wittgenstein it cannot even be expressed. Novelty connects both these ideas. Novelty, as the connection between this temporality and subjectivity, what sets our consciousness apart from the rest of nature.

The boundary separating the describable from the ineffable changes over time, due to the reality of novelty

Suggestion (3) ***The boundary between "that whereof we cannot speak" and that which is describable changes as a function of time, due to the reality of novelty.***Perhaps the most provocative and poetic aspect of Wittgenstein's early thought is his reference to the inexpressible, and his emphasis on its importance. Wittgenstein identified the boundary between expressibility and "*that whereof we cannot speak*" as homologous with the boundary separating the describable from what he called the mystical, *"the feeling of the world as a limited whole."* This, by definition, cannot be described, and (more tentatively) can perhaps only be experienced subjectively and "intuitively." If this premise is accepted, a problem occurs regarding the nature and definition of linguistic nonsense. A traditional reading of the *Tractatus* has interpreted Wittgenstein as meaning that somewhere underneath what is, technically speaking, linguistic nonsense, there is, sometimes, something inexpressible but significant. Areas of human thought such as ethics and aesthetics are categorized into this inexpressible field of (intuitive?) understanding. The most important part of the *Tractatus* is the part that Wittgenstein left out, as he himself said. Importantly, if accepted, this perspective would imply that there exist two fundamentally different kinds of nonsense. First, there is mere gibberish, second, something profound but unsayable. This principle underlies the labeling system above, in which subjectivity is not reducible to mere gibberish, subjectivity is deeper and richer, and cannot be dismissed by the analysis of language with respect to grammatical or syntactical criteria. We do not end up with a "sense of the mystical" just by rearranging words, and deciding whether or not statements are nonsensical. There is a qualitative difference between inexpressible "artistic", emotional, subjective sentiments, and propositions that are merely linguistically incorrect. But the key issue connected to this traditional reading of the *Tractatus* is that of whether any such "sense of the mystical" can be said to be a coherent concept at all, let alone be said to exist.

We might take a skeptical approach to a person who claimed that they had had a "mystical" or "spiritual" experience or feeling, in response to (say) a stimulus of natural beauty, for instance a beautiful sunset. We would probably begin by asking them precisely what they meant. If one were a fervent "scientific" materialist, one might explain to the person who claimed a link with "the mystical", that feelings of emotional enjoyment were the consequence of some sort of connection within our evolutionary past, between pattern in the world (either beneficial to our survival and therefore positively interpreted, or exhibiting some form of emergent "beauty") and the functioning of neuro-transmitters in our brain. Such a person might say something along the lines of "you found the sunset beautiful because it triggered the release of (say) dopamine in your brain. This release was (say) selected for in the evolutionary past, because the sun, as a source of warmth and light, is beneficial to survival, and therefore neurological responses that promote increased association with sunlit locations increased by virtue of natural selection." This is the kind of perspective by which a skeptical person might (to some extent reasonably) argue that there is simply nothing that needs explaining when someone refers to an experience that is "mystical" or "beyond words." This sort of natural-selection based "just so story" is, as I have caricatured it, perhaps a little dismissive, but it may well be true. Feelings of beauty, or subjective emotional experiences, or anything else that someone might designate as being indescribable in words, might very well have some sort of atomist, materialist, negationist interpretation. A "spiritual"/"mystical"/"indescribable" experience in response to a beautiful sunset, during a time at which a particular person is particularly emotionally receptive, may be "just" dopamine or serotonin, "just" recognition of an emergent natural phenomenon, "just" a by product of adaptation in our hunter gatherer past, etc. The experience is "just" a product of these external explanatory criteria, there is no need for any other "deeper" explanation. There is certainly no need, from this skeptical perspective, to start saying "you saw God in the beauty of the sunset", or some analogous caricature of how a religious person might interpret (misinterpret?) the fact that Wittgenstein throws around phrases like "the mystical."

But conversely, at the most direct phenomenological level, language should reflect the sensations experienced within the consciousness of individuals. If an individual has a sense of experiencing a pattern, a sensation, a perspective, which warrants explanation, the role of language should be to express that. This expression may be best achieved either by pre-existing or as yet unevolved linguistic conventions. In other words, regardless as to whether a sensation of "the world as a limited whole" or equivalent "mystical" stimulus, is "real" according to prescribed (scientific, logical, linguistic) criteria, or is not, the existence of such a sensation is a "real" feature of consciousness. From this point of view, the feeling that such sensations justify expression in language, and are therefore more than mere syntactic nonsense, is equally real, even if attempts at such an expression turn out to be unsuccessful. The key point is that the limits of language define more than just the rules of language, they define "that whereof we cannot speak" as a real category. In this second perspective, the sensation of something important but inexpressible is not the same thing as grammatically/syntactically incorrect or poorly defined propositions, because such propositions are not associated with such a subjective sensation.

The "resolute reading"[41] of the *Tractatus* takes the first, skeptical position to its full conclusion, and challenges the traditional interpretation of Wittgenstein as invoking some mystical/inexpressible category of human experience. Its originality and internal consistency is controversial[42], but the clearest aspect of the resolute reading involves the argument that there is only sense or nonsense. There is no special ineffable category that is surficially linguistic nonsense, but actually something deeper. A resolute reading is, as McManus[43] puts it "*a commitment to reading the book as not defending a particular metaphysics and a particular philosophy of mind and language.*" Furthermore, resolute readers of the *Tractatus* argue that

> "*The confusion that the Tractatus seeks to dispel is not a confused theory, such that a correct theory would be a proper way to clear the confusion, rather the need of any such theory is confused. The method of the Tractatus is to make the reader aware of the logic of our language as he is already familiar with it, and the effect of thereby dispelling the need for a theoretical account of the logic of our language spreads to all other areas of philosophy. Thereby the confusion involved in putting forward e.g. ethical and metaphysical theories is cleared in the same coup.*"

In other words, the purpose of the *Tractatus* is, as Wittgenstein says, to demonstrate the limits to linguistic expression that the user of a logical language is already aware of. Most significantly, within the resolute reading, when the early Wittgenstein states that the "ladder" of the *Tractatus* can be kicked away after one has climbed up it, that "he who understands me recognizes them as nonsensical", he means that the work is, quite literally, nonsense. The exercise of climbing the linguistic ladder is to expose it as having a logic of which we are already aware. By analyzing the structure of this ladder, and speaking in terms of "language world conformity", we are becoming wrapped up in confused pseudo-problems, analogous to asking "why is it always the winner of a race who occupies first place?"[18]. The very notion of a "general propositional form" for language, "homologous" or "intrinsically similar" to the structure of the world it is describing, and the idea that it is significant or interesting that such a form exists or has such a structure, is, resolutely, dismissed as deriving from conceptually confused questions. Language is "structured" like the world because if it is not, it is not language- merely marks on paper, or incoherent sounds.

Consistently with such a view, McManus interprets the *Tractatus* through a basic principle that he calls "sign symbol confusion." Wittgenstein differentiates between "signs" (contextless shapes, letters, sounds) and "symbols" (appropriately contextualized shapes, letters and sounds that have a consistent meaning through linguistic conventions), and notes that the idea of "discovering" intelligible use of language by achieving conformity between the language and the world is nothing more that correcting an initial, erroneous confusion between a decontextualized (therefore unintelligible) sign, and an intelligible symbol, through which language-world conformity occurs.

> "*When we imagine the intelligibility of a name as lying in its 'conformity' with the object it denotes, we have indeed failed to assign sense to the 'possibilities' that we imagine here as the name's being used intelligibly and its being used unintelligibly. The illusion that we have got hold of a substantial issue here arises out of another sign/symbol conflation...*"

Importantly, this perspective emphasizes that Wittgenstein's cryptic remarks about things like internal similarities, the mystical, and most specifically seeing the nonsense of his own claims and kicking away the ladder, are meant to show that, above all, the language that might be used to point at them is confused:

> "*Wittgenstein works through the confused pseudo-logic of our conformist commitments, and the 'claims' that he makes elaborate on, extend, the nonsense that those confusions prompt us to talk; thus Wittgenstein's own claims are, as $TLP6.54$ indicates, nonsensical.......to understand Wittgenstein's talk of 'internal properties', 'internal relations', 'the inexpressible', 'showing' and 'the transcendental' is to recognize that the thinking that leads us to it is informed by confusion. The 'insights' that this talk expresses bring us to the point of recognizing and being able to reject confusions within which theses 'insights' emerge. What one is left with is an understanding of the nonsense that one previously unwittingly talked.*"

McManus goes on to emphasize how stating that there exists some special, unsayable ethics or aesthetics, is an example of this sign-symbol confusion:

> "*The 'sayable' that we come to feel must be supplemented by an ineffable moral knowledge is a confused hybrid of sign and symbol, supporting and supported by a confused vision of expression......the 'recognition' of the 'need' for inexpressible and unteachable powers of 'conscience' emerges for those in the grip of this confusion; when we shed this confusion, we ought also to shed our belief in those 'powers' as solutions to (what we now see are unreal) problems. Another image of a problematic gap between subject and world which we come to believe must be spanned by a perplexing insight turns out to be the product of sign/symbol conflation, of our enchanting and being enchanted by mere words.*"

Within the resolute reading, Wittgenstein's goal is therefore not to defend any particular metaphysics (inexpressible or not) with his comments about things like "the mystical", but rather to simply show us that consideration of such things is a violation of the logic of our language, a logic about which there is nothing mysterious or unfamiliar. Obviously the structure of our language conforms to that of the world, because it cannot be any other way

- that's what meaningful language is. The "general propositional form" is just there, obviously, because that it how our language necessarily works - it has no specific metaphysical implications. Understanding the intelligibility (of language) as conformity (with the world) is misguided, because it is impossible to imagine matters any other way, because language that does not conform to the world is not language. The resolute reading takes this attitude to its full potential conclusion; there is no difference between gibberish-like unintelligible sounds and (say) trying to define objects and states of affairs with reference to something else.

The resolute reading is complex and nuanced subject way beyond my own expertise, so I will not attempt to critique it in any detail here. But two key claims arising from it appear relevant; first, that when Wittgenstein says the *Tractatus* is nonsense, he means it in a quite literal way, and second, there is no metaphysically significant or special category concerning the "unsayable", the ineffable. In my opinion both these claims are incorrect, for reasons that bear directly upon novelty. As numerous professional philosophers have commented, even if the *Tractatus* is, in some sense, with respect to some sort of external criteria, nonsense, it is still a particularly elaborate and engaging type of nonsense. Reading *Tractatus Logico Philosophicus*, even if you later decide it to be ultimately nonsensical, is a very different experience from (say) listening to the shouting of a man who is so drunk that the sounds of his voice are not even discernible as words. In other words, there is, as discussed above, a fundamental difference between nonsense (mis-application of language, sign-symbol confusion etc) and gibberish (non-language, lacking in any tangible connection to thought). More importantly, as I interpret Wittgenstein, when he describes the *Tractatus* as nonsense, he means not that it is gibberish, but that it is what Braver calls groundless. The logic of the *Tractatus* is internally consistent, and there is such a thing as language-world homology, but the foundations of this consistency and this homology come to an end, meaning that the book's logic is finite. For instance, if (for argument's sake) it were possible to empirically/experientially develop a watertight definition of an "object" (for instance if the "point mass" or some other indivisible physical entity were found to exist), then the *Tractatus* might well make perfect sense. Problems begin when one questions the foundations, by asking Wittgenstein "What do you actually mean by an object?", "Can you give a specific example of a state of affairs?" etc. But if we just arbitrarily designate particular types of object as being "that which combines in particular types of states of affairs", and accept the shaky groundlessness of the foundation, then the *Tractatus* becomes almost a tautological truism. The structure of the proposition must have some sort of "internal similarity" with the structure of the fact it describes, if objects really did exist, the general propositional form would be recovered as valid. The difference between the *Tractatus* and *Philosophical Investigations* is that in the latter there is an appreciation that essentially anything can be designated as an "object", to be combined in any sort of state of affairs - i.e. an appreciation that one can build the foundations of meaningful language anywhere, so long as a language game can be played. The *Tractatus*'s generality becomes shaky due to the inherent groundlessness of all language, and needs to be modified to be more contextual and fluid. But this does not mean that the *Tractatus* belongs in the same category as gibberish.

A further problem with the resolute reading arises in relation to the question of where meaningful language is supposed to come from in the first place. I suggest that because the reality of novelty causes "meaning as use" to evolve over time, a (pseudo) linguistic convention that was nonsensical or confused at one point in time can become sensical, conducive to the development of a language game, at a later point in time. For any given word or syntactical/grammatical convention, a time necessarily must have existed at which it was invented, based on even the most crude awareness of human evolution. At a point in time, any given word must have been nonsense, and it must have only ceased to be nonsense once a coherent (and new) language game became established. The "correction" of attempts to step outside the logic of our language, which appears to be advocated by the resolute reading of the *Tractatus*, only works if that logic is fixed. If the logic of our language evolves over time, so too must the recipe for ensuring that propositions adhere to that logic. If the world and the conventions that are useful in our attempts to describe it were fixed, then even a groundless language system could be general. If the world were a single entity with universally consistent rules, fixed and consistent, one logically perfect language could suffice to describe it (or at least to describe the describable proportion of it). But this is not what our language is. We have multiple different contextualized systems of meaningfulness, because the world is heterogeneous in space, but also heterogeneous in time- the latter property being a consequence of novelty. Language games have a family resemblance because they evolve temporally. They evolve temporally because novelty changes the foundation of meaningful language-world homology upon which they are built.

Take the resolute perspective at face value - there is only one kind of nonsense, the *Tractatus*' ladder is indistinguishable from gibberish, and there is nothing mysterious or metaphysically significant about language-world homology, it simply reflects how language has to be. But now suppose that the world changes, so that language that was meaningful at the start of this evolutionary interval is rendered meaningless, and/or elements of the world that can potentially be meaningfully described have not yet been tackled by language, because they are new. To render the newness in the world describable, and to clear up the meaningfulness of the now redundant aspects of linguistic rules, we need to work through the nonsense/meaningfulness boundary. We need to develop new words for the world-novelty we encounter, and acknowledge that some of the linguistic rules that we were previously using are now redundant. To do this, we have to use the language that was previously meaningful but is now nonsense, and come to realize its change in status. If this change instantly lumps the redundant features of language into the same category as complete gibberish (as in a resolute reading), then we are required to just instantly recognize this, and somehow intuit that the natural logic of our language has moved on, pick up on this, and keep quiet about the now obsolete rules that we were previously using. We are actively discouraged from stepping outside the logic of our language. However, if we are allowed a more fluid, semi-meaningful status for certain linguistic propositions then it is much easier for us to imagine the evolution within time of meaningfulness in our language games, through the back and forth of public communication and semi-meaningful language-world relations. The resolute reading of the *Tractatus*, because it is at root too puritanical about the designation of nonsense, resists this evolution - and thereby gives rise to an unrealistically static notion of language that is incompatible with the incursion of novelty into language.

The idea that there is no special category of the unsayable, no metaphysical, ethical or spiritual significance to "that whereof we cannot speak", nothing important about the ineffable (other that its use as a tool to remind us of the the logic of our language), is also near impossible to reconcile with the fact that Wittgenstein explicitly stated that the part of the book he had defined by leaving out was the most important[19]. The *Tractatus* does not defend a *particular* view of the ineffable, or the mystical, but it does defend the existence and importance of both these categories. The connection between the ineffable and ethics/aesthetics might be taken to mean that ethical judgments ultimately fall back (i.e. given enough sequential attempts to justify an ethical choice in a value system) upon an emotional/intuitive/aesthetic "choice" or "feeling." There, by definition, not a great deal to say about this. The key point is that it is not possible to develop a particular view about the ineffable, because particularity requires specification within linguistic (i.e. "effable") criteria. But the fact that the ineffable is not particular does not mean that it is not significant.

A further consideration arises regarding the boundary separating the describable from the indescribable. The position of this boundary changes over time, due to the reality of novelty. If the sphere of experience that is coherently, objectively communicable within language is a category that has a degree of fluidity to its boundaries, then so too must the category of "what is mystical." Most obviously, modern science and engineering can explain phenomena (say physical fields such as magnetism) that in early times may have been deliberately dismissed as "mystical" or "magic." The natural fluidity inbuilt into the boundary between expressibility and subjectivity means that, as is well appreciated, one age's mysticism is another age's mundane. Assuming any given society has a memory of previous technological innovation, and that linguistic expressibility necessarily goes hand in hand with this innovative history, then there is a progress that naturally pushes back the boundary of what is beyond expressibility. The size of "the mystical" as a category might well be argued to decrease over time. The obvious science-centric perspective is that there is much less now that is mystical than there was in stone age times. But a religiously inclined person might argue that "God" injects comprehensibility into the world by intervening in our experience and actively revealing aspects of it. A materialist-atheist motivated by scientific progress might argue that we use our intelligence and mental innovation to conquer nature. The difference is a matter of taste, and is to a large extent unknowable. Furthermore, there is no reason to suppose that the "size" of the total sphere of experience that is either comprehensible and expressible (i.e. the sum of the expressible and the mystical/ineffable) is fixed, or has any consistent relationship to time. For argument's sake, there may be more to potentially express at one point in time than at another.

But all this is not really what Wittgenstein was getting at when he described the "feeling of the world as a limited whole", or things which "make themselves manifest." It seems as if his intention in writing the *Tractatus* was to emphasis the contingency, the Groundlessness[25] of the logic of language and how it manifests at the boundary of expressibility. The clearest way to see that this is to note what this mystical feeling actually is within human consciousness. Is there really a fundamental difference between an ancient palaeolithic man feeling a sense of profound contingency about why the sun rises or where the seasons come from, and a modern man feeling an equivalent sense of contingency when running up against the limits of scientific knowledge (say, the question of what caused the origin of life, or what "existed" before the big bang)? Culturally, modern men may choose to deal with unanswered questions as either an object of faith - either in "God" or in the capacity of science to one day answer the question, but, crucially, the conscious, subjective sense of contingency that Wittgenstein referred to as the mystical is the same. It is only the reference point, and the conventions of both society and language, which have obviously changed between hunter-gatherer times and post industrial secular society. It is the subjective feeling of the inherent limits, to comprehensibility, to the ineffability/expressibility boundary, to the extent to which the contingency of our existence can be known - to which Wittgenstein (as I read him) refers. This is not the same thing as mere demarcation of the boundary separating what is known from what has yet to be discovered. This kind of contingency is, of course, exactly what Heidegger seems to have been attempting to reflect by posing the (meaningless) question "What is the meaning of being?" The clearest way to tie together the work of Wittgenstein and Heidegger is to blend the most cryptic assertions of each thinker, and argue that the meaning of being is what is mystical.

Novelty changes the systems in the world (changing what can potentially be expressed) and the way that consciousness responds to the systems with which it is presented (changing what might evoke feelings of inexpressibility and/or the mystical). Novelty changes the distribution within time of causal rules connecting systems of matter within space. This changes the patterns that will appear in the world as a function of time, include the patterns that might evoke the feeling of the world as a limited whole. Things exist in the world that previously did not, due to the reality of novelty. One might say that novelty makes itself manifest. Novelty also changes the subjective dispositions and objective ideas within consciousness. This change in the "wiring" of consciousness (as opposed to the wiring of the brain), such that conscious responses to any stimulus will differ, including potentially the responses that elicit feelings of that whereof we cannot speak. More significantly, once the consciousness has access to a new idea it has a new way of thinking about the world that may allow it to comprehend what was previously incomprehensible, describe what was previously indescribable. Thus, both in the world and internally to consciousness, novelty gives a temporality in the boundary between what is, and is not, mystical.

This argument has bearing on the relationship between novelty, science and art. I suggest that art is, by definition, the instinct that distinguishes the pattern that warrants an explanation from the pattern that does not. Art can therefore describe novelty directly (but not explain it). Science or other logical inductive systems of thought can only subsequently describe the repetition that follows novelty's initial occurrence. We can express a feeling for the mystical in art, but we cannot describe or explain this feeling, or the stimulus for it, in objectively meaningful language. Art is therefore the place where we deal with novelty directly and at face value, science and reason where we deal with the repetition that novelty subsequently gives rise to. This mutual exclusivity is why science, logic, inductive reasoning, and other attempts at the development of meaningful languages leave us dissatisfied when describing a stimulus that promotes a feeling of something profound but inexpressible within our subjectivity. For instance, suppose we see a sunset, which provokes in our subjectivity a feeling of beauty. Some individuals might develop religious or faith-linked emotions or intuitions during this experience, some feelings of "nature worship." Some individuals might experience little from the sunset. Some individuals with a materialist/science-centric worldview might find that emotional experience of beauty evokes feelings of religious/spiritual mysticism, and suppress such feelings as irrational, or attribute them to some previous evolutionary advantage to pattern recognition. But regardless of the variation between individuals in how they respond to the sunset, nobody can quite put into words exactly what it is to experience a beautiful sunset. Obviously, in a literal sense, there are many ways to make the statement "the sunset is beautiful." We describe the sunset as (say) the refraction of the light as it passes through layers of the atmosphere with differing constituents, or the release of serotonin in our brains in response to recognition of a environmental factor that positively influenced our ancestors' survival. These are by no means meaningless statements, but the meaningfulness of this level of explanation is far removed from the subjective experience we started out trying to describe. And it is impossible to develop a meaningful language to describe the subjective experience of being deeply emotionally moved, to the point of a feeling of what Wittgenstein called mysticism, by a beautiful sunset or any other stimulus evoking feelings of the contingency of existence. This disconnect arises precisely because meaningful expression and the emotional feeling of "mystical" contingency are mutually incompatible. Art is not meaningful language, because of its lack of consistency and objectivity, science in some sense is, the difference being that science has rules. Perhaps this issue of the problem of expressibility may perhaps be part of the reason why philosophers have frequently concluded that art, not science, is the deepest means of expressing what human existence is.

Wittgenstein was correct that "that whereof we cannot speak" is more fundamental than what can be said meaningfully, and was "the mystical." But he was incorrect about the finality with which he (initially) attempted to delineate this boundary. Novelty happens at this boundary, after an instance of novelty, something can be described meaningfully when previously it could not have been. Therefore, the boundary between the mystical and the describable itself evolves over time, because novelty effectively adds things to the describable world that were not previously there, and adds a predisposition to our consciousness, both to describe some patterns and to attribute the need for a description to others. Both these effects have the consequence that novelty changes the position of the boundary between that whereof we can, and cannot, speak meaningfully.

Novelty explains why there is an aspect of disclosure to truth

Suggestion (4) *The need for a truth concept based on disclosure, as opposed to convergence/correspondence, derives from the introduction of new truths to the world by instances of novelty.* Novelty underlies the specialness of time that Heidegger picks up on throughout his philosophy. Heidegger's methodology involves classifying what are ultimately uniquely human aspects of being and opposing them to the rest of existence, then elaborating upon these uniquely human qualities in a manner that emphasizes a temporality and an interactivity. All of the core concepts that distinguish Daesin from ready to hand or present at hand objects, or distinguish authentic from inauthentic modes of being, exhibit, as Heidegger defines them, what might be called an "enhanced temporality." For instance, the basis of human being in the world, Daesin, is care. The meaning of care is temporality. So the foundational basis of the definitive aspects of human being in the world is "made of", or at least "based in" care, and the meaning of care is some sort of "time-ness", which is absent from that which is not care. Thrownness, fallenness, immersion in the they-self and any other modes of being that do not genuinely reflect what we are, can be distinguished from authentic modes of being, in that the latter are more genuinely grounded within time. When we adopt an authentic mode of being we are aware of the call of conscience, reminding us of the inevitability of our own death. When we are authentic we are "more" in time because we are more towards death, through a "pulling" that operates temporally. We are aware of our innate predisposition to "historicize", i.e. we are reminded of our historicality. Authenticity is, in some sense, more temporal than inauthenticity, and this is why authenticity more directly reflects what we are. In a moment of vision, we confront our anxiety about death, and our guilt about inauthenticity, and become aware of something foundational to our being. The moment of vision is a psychological slap in the face with the inherent temporal-ness of existence, a conscious reminder of the foundational importance of time.

A further emphasis on time becomes apparent in the context of Aletheia (unconcealing, disclosure), and the deep link that Heidegger proposed between disclosure and the truth. Virtually all other philosophical conceptions of truth are in some way based on correspondence or convergence[44]. In more conventional truth concepts a statement about the world is true if it corresponds to how things are. Similarly, two different statements or observations are true with respect to the criteria established by the other if they converge or agree, if they exhibit some sort of homology between each other. This is a fundamentally different kind of idea to aletheia. Aletheia is closer to revelation, to the unveiling of something that was masked - one might even say, with regard to the above discussion of Wittgenstein, that aletheia amounts to the rendering comprehensible of that which is previously mystical. Although Heidegger later acknowledged that the unveiling/disclosure of aletheia need not necessarily be the same thing as truth[4] , he retained the idea that aletheia was a fundamentally important aspect of meaning. The difference between aletheia and convergence, correspondence, or other more conventional conceptions of the truth is, once again, an emphasis on temporality. If we converge on the truth, this means we discover the same thing in different ways. The notion of aletheia, truth as disclosure, is still consistent with correspondence and convergence being ways in which we can access truth, but adds the idea that the truth is revealed to us in disclosive events of "unveiling." If something is disclosed to us, it is disclosed over a time interval, at the beginning of which it was hidden, and by the end of which it is manifest. Such disclosure is, of course, discontinuity in the distribution, over time, of what is comprehensible, a change in the boundary between the ineffable/mystical and the expressible: a novelty. Equivalently, the moment of vision is a sudden grounding of our consciousness in the inherent temporality of existence, a reminder of the inevitability of our death, and a change in perspective through this grounding. We therefore have two of Heidegger's core concepts, aletheia in the world, the moment of vision in our consciousness.

Both these concepts are inherently temporal, and both relate to time in a dramatic, comparatively discrete way, much like novelty as described here. Thus, Heidegger's work is conceptually strengthened by an appreciation of the reality of the new:

Aletheia is novelty in the world and/or consciousness

Aletheia is novelty. Truth is closer to disclosure than correspondence, because what is true and what is not changes over time, due to the reality of newness. During non-novel intervals (of repetitive causation and/or consciousness), truth as correspondence is a fully coherent concept and need not necessarily be inconsistent with the reality of aletheia. Truth as disclosure gives a directionality to time that, while difficult to reconcile with truth as convergence, is fully consistent with the relationship between time and the new. Novelty is what makes truth disclosure-like rather than correspondence-like, and what creates the tie between truth and the directionality of time. If we converge on the truth from different directions that turn out to have similarities, the truth could easily have been there, fixed, for the whole time, waiting to be discovered. If the truth changes over time, it is possible that two different ways of investigating what is true might agree at one point in time and disagree at another. This kind of changing over time, if it holds, becomes problematic in terms of the notion of truth as convergence, because what is there to be converged upon is itself subject to change. Aletheia, or another more dynamic truth concept, does not necessarily suffer from this kind of problem, because it is inherently tied to a discrete point in time at which the unveiling happens (much like novelty). However, aletheia cannot be a complete concept of truth. After the disclosure event, one needs a correspondence truth concept (or at least some other reference point) in order to determine whether what was previously undergoing disclosure, but is now simply present, is true. Therefore one can say that aletheia adds to what is true, whereas correspondence (between theory and reality) dictates what is true. Aletheia introduces new truths to the world. This is exactly analogous to how novelty introduces new repetition to the world - in that the two different concepts require each other in order to produce a coherent picture.

If the truth is revealed to us in discrete events of disclosure, it is also possible that it was there all along, but it is necessary to ask "What does the disclosing?" What is the nature of the process or influence that causes an undisclosed truth to become disclosed? Is this process part of the truth, or something distinct from the truth? The process of truth disclosure cannot, by definition, be fixed, because it is embedded in time, one obviously cannot have atemporal disclosure. So if truth disclosure is part of the truth, i.e. an aspect of it, the truth is not fixed, the truth is temporal. If this disclosive aspect is not part of the truth, then there is no way (without correspondence or other temporally consistent truth criteria) to know that what is disclosed is genuinely true (it might be illusory). A simple way out of this semantic mess is to say that an instance of aletheia is an instance of novelty, at which something becomes true that previously was not true. Then the unveiling event is part of the truth and non-illusory, and things are true because our theories agree with the world and because different theories converge on the same thing. New truths come into existence during instances of novelty. The call of our conscience to authenticity during a moment of vision is also arguably closely related to the subjective experience of having a novel thought. The rearrangement of perspective, aswell as inherent time dependence and subjectivity, creates a direct tie between the two concepts. However the parallel is not quite so strong - for instance Heidegger never specifies whether it is possible for us to experience the same moment of vision twice (if this were possible it would rule out the tie with novelty). Furthermore, aletheia could be unveiling within consciousness just as much as within the world. The key parallel between aletheia and novelty is the addition of the directionality to time.

Meaning and subjectivity are temporal, due to the reality of novelty

I have suggested here that appreciating the reality of novelty, within our subjectivity and within the world, provides a way to see the consistency that is present between the work of Wittgenstein and Heidegger. Novelty is the underlying reason for the unavoidable temporal evolution of the logic of our language. Novelty is what separates the linguistically expressible from the unavoidably ineffable. Novelty separates that whereof we cannot speak, from that of which we can speak, and speak clearly. Novelty makes the boundary that defines this separation inherently temporal, raising a direct parallel to Heidegger's conception of truth. Novelty is also what makes our conscious existence temporal: it is through the entry of subjective novelty into our consciousness that we recognize the recognize the foundational "time-ness" of what we are. Therefore, novelty provides a kind of lynchpin through which Heidegger's recognition that we are somehow "embodied time" and his introduction of a time-dependence to truth, can be directly related to Wittgenstein's change of mind concerning the finality of the *Tractatus*, or any other single, constant, logical linguistic characterization of the world.

Languages need to be contextual and need to evolve for the same reason that we are temporal beings who possess attributes such as historicality - the reality of the new. If Wittgenstein's initial intuition that "the mystical" is demarcated by the limits of linguistic description, then the reality of novelty implies that the boundary between the mystical and the non-mystical itself evolves over time. The feeling of the world as a limited whole, and the mysticism it evokes, may, for now, be a consistent feature of human consciousness. But the patterns in the world and in consciousness that stimulate such mysticism are just as temporal as any other feature of our existence. The boundary between art and science (some might say between the spiritual and the practical) is itself not fixed. The rules of any language have evolved prior to a particular instance of novelty, therefore these rules must subsequently adapt before any given novelty can be described with consistency. Meaning as use evolves in time, as Wittgenstein noted, because we, and therefore our language, are inherently temporal, as Heidegger noted. The cause of this unavoidable evolution is the reality of the new.

Notes

1. Paul Strathern. *Heidegger in 90 minutes.* Ivan R. Dee. 2002.
2. Searle, J. *The Word Turned Upside Down*, review of *On Deconstruction: Theory and Criticism after Structuralism*, The New York Review of Books. 1983.
3. Grayling, A.. *Wittgenstein: A very short introduction.* Oxford University Press. 1996.
4. Inwood, M. *Heidegger: A very short introduction.* Oxford University Press. 2000.
5. Russell, R. *The Principles of Mathematics.* Norton. 1938. (First published 1903).
6. Zalta, Edward N., *Gottlob Frege* Stanford Encyclopedia of Philosophy http://plato.stanford.edu
7. Strathern. *Ibid.* ref. 1.
8. Kant, I. *Critique of pure reason.* (Cambridge, 1999. (First published 1781)).
9. Georges, R. *The Analytic/Synthetic Distinction.* Stanford Encyclopedia of Philosophy 2014 http://plato.stanford.edu
10. Bruce, R. *A Priori Justification and Knowledge.* Stanford Encyclopedia of Philosophy2014 http://plato.stanford.edu
11. Creath, R. *Logical Empiricism*" Stanford Encyclopedia of Philosophy 2014 http://plato.stanford.edu
12. Schlipp, P. *The Philosophy of Rudolf Carnap.* Open Court Publishing Company 1999.
13. Monk, R. *Ludwig Wittgenstein: The duty of genius.* Penguin Books. 1991.
14. Inwood, *Ibid.* ref. 4.
15. Wittgenstein, H. & Pascal, F. *Recollections of Wittgenstein.* (Oxford, 1984).
16. Edmonds, D. & Eidinow, J. *Wittgenstein's poker: The story of a ten minute argument between two great philosophers.* (Harper Perrenial, 2002).
17. Inwood, *Ibid.* ref. 4.
18. Wittgenstein, L. *Tractatus Logico Philosophicus.* (Annalen der Naturphilosophie, 1918 (1921)).
19. Wittgenstein, L. *Philosophical investigations.* (Prentice Hall, 1953).
20. Wittgenstein, L. Informal comment in a letter to Ludwig von Ficker, 1919.
21. Waddington, C. *The ethical animal.* George Allen & Unwin. 1960.
22. Wittgenstein, L. *Remarks on the Foundations of Mathematics.* (Wiley Blackwell, 1981(First published 1953)).
23. Bernays, P. *Comments on Ludwig Wittgenstein's Remarks on the Foundations of Mathematics.* Ratio, 2 (1): 1–22. 1959.
24. Roydich V, *Wittgenstein's Philosophy of Mathematics.* Stanford Encyclopedia of Philosophy. 2011 http://plato.stanford.edu
25. Godel, K. *On Formally Undecidable Propositions of Principia Mathematica and Related Systems.* Monatshefte für Mathematik. 1931.
26. Smullyan, R. & Fitting, M. *Set Theory and the Continuum Problem.* Dover. 2010.
27. Wittgenstein, L. *Zettel.* (Wiley Blackwell, 1981(First published 1968)).
28. Wittgenstein, L. *On certainty.* (Harper and Row, 1953).
29. Moore, G.E. *A defence of common sense.* Contemporary British Philosophy (2nd Series) 1925.
30. Heidegger, M. *Being and Time.* (Blackwell 1962 (First published 1927)).
31. Russell, B. *Wisdom of the West.* (Crescent Books, 1989).

32. Braver, L. *Groundless grounds: A study of Wittgenstein & Heidegger.* (MIT, 2012).
33. Wiegand Petzet, H. *Encounters and Dialogues with Martin Heidegger, 1929-1976.* (Chicago, 1993).
34. Huffman C. *Pythagoreanism* Stanford Encyclopedia of Philosophy 2015 http://plato.stanford.edu
35. Monk. Ibid. ref 13.
36. Cavell, S. *The claim of reason.* (Oxford, 1979).
37. Dreyfus, H. *What Computers Can't Do: The limits of artificial intelligence.* (Harper-Collins, 1978).
38. Dreyfus, H. *What computers still can't do: A critique of artificial reason.* (MIT, 1992).
39. Hintikka, J. *Wittgenstein on Being and Time.* Ludwig Wittgenstein: Half truths and one and a half truths. 241 – 247. 1996.
40. Chomsky, N. *Three Models for the Description of Language.* IRE Transactions on Information Theory 2 (3): 113–124. 1956.
41. Crary, A. & Read, R. *The new Wittgenstein.* Routledge. 2000.
42. Hintikka, J. *What does the Wittgensteinian inexpressible express?* The Harvard Review of Philosophy *XI*. 2003.
43. McManus, D. *The enchantment of words: Wittgenstein's Tractatus Logico Philosophicus.* (Oxford, 2006).
44. Glanzberg, M. *Truth.* Stanford Encyclopedia of Philosophy2014 http://plato.stanford.edu

CONCLUSION: AN UNANSWERABLE QUESTION

It has been argued throughout that novelty results when systems that function in similar ways but are made of different stuff happen to interact. In terms of the nuts and bolts of how different spatial arrangements of matter interact with each other, a key conceptual distinction is that between structure and function. With respect to a given system (a given contextualized structure-function relation), function is local repetition of causal rules, structure is global repetition of causal rules. Structure constrains function, not the other way around. Novelty is not only a symbiosis between recurrence and recombination, as has traditionally been assumed. Novelty is a symbiosis between systems that are on the one hand functionally similar (such that the cause-effect patterns of their functional repetition fall into synchrony with each other), but on the other hand structurally different. The combination of the homology in the function-function relationship, and the fact that the function of each system is, in effect, trying to deal with the alien nature of the structure of the other system, leads to the random, transient repetition of associations between particular changes. In some interactions, these transient associations develop into full blown repetitive causal rules, in association with the emergence of a structure-function relationship at the level of both systems. Because this relationship includes new causal rules (consistent, repetitive associations between particular changes, that did not exist prior to the initial interaction) it constitutes qualitative novelty in the kind of cause and effect that happens within matter and space over time. In other words, nature is inherently symbiotic, and a novelty is a symbiosis between structurally different and functionally similar systems. Life is the most novel natural phenomenon, mind is the most novel aspect of life, human minds appear to be the most novel kinds of minds (because the human mind is effectively a "symbiosis" between ineffable subjectivity, and linguistically representable objectivity). The history of life on Earth is characterized by symbiotic recombination, and the combination of extreme environments and equally extreme instances of cooperative biological reorganization.

In terms of internal human subjectivity, a instance of novelty is the origin of a new idea or a new consistent association between a particular consciousness and some aspect of the world. Novelty is inherent to human subjectivity, and time is the medium within which novelty becomes relevant to it. The truth is habitual, not fixed, and an objective instance of novelty constitutes the disclosure of a new truth. In Wittgenstein's terms, an instance of novelty is the origin of a meaningful language-world relationship that did not previously exist, a shift in the position of the boundary between the ineffable and the describable. In Heidegger's terms, an instance of novelty corresponds to an act of the unveiling of a truth, a moment of vision is very similar to the psychological experience of novelty, and novelty is what gives a directionality to time - what makes Daesin a fundamentally temporal entity. But the impact of novelty upon our subjectivity also shows us how sometimes the more intuitive aspects of our consciousness step outside the logic of our language. Sometimes, this gives rise to new, logically representable thoughts that describe the world in meaningful sense. Meaningful language starts out as a germ of subjectivity; a situation in which we feel compelled to say something, we intuit that there is something to say, but we "cannot find the words" within the constraints of meaningful language. This, I suggest, is at least part of the reason why Wittgenstein changed his mind about the fixed notion of a general propositional form, in favour of a dynamic, contextual, and temporally evolving notion of meaningful language. A fixed "general propositional form" exhibiting a constant "language world homology" would be unable to flow with the novelty that we encounter in the world and in our subjectivity, unable to evolve over time. Only evolving, fluid, contextualized language games can sustain linguistic objectivity in the face of the sort of inherent temporality of human existence that was emphasized by Heidegger. Language evolves because what can be meaningfully described also evolves, as does what we want to say, and have the capacity to express to others. Meaningful language is a temporal concept, because novelty is real.

The reason that nature has within it the disposition to cause disparate systems to come together and produce symbiotic novelty cannot be described in the language of how such systems work individually. The causation relevant to the internal function of such systems is completely uncorrelated to the causation that initially brings these systems together. This is why novelty can only be identified retrospectively, rather than predicted in a strict sense. More simply, on some level it is "just luck" that systems that function in similar ways but are made from different stuff happen come together in time and space in such a way as to produce novelty. This isn't much of an explanation, and we must ultimately fall back on the anthropic principle: If novelty wasn't real, the prerequisite increases in complexity necessary to produce observers who ask questions about novelty would never have occurred. In order for us to be here asking questions about novelty, novelty must exist. Like most arguments based on anthropic reasoning, this is all well and good; we must content ourselves with changing a "why?" question to a "how" question. The question becomes "How does nature produce novelty?", we know better than to ask "Why?" But in terms of individual discontinuities in the distribution of cause and effect within time, novelty seems undirected, a chaotic breaking away from what went before. By anthropic principle-derived argument games, we can convince ourselves that novelty exists because it must. But if we look at what novelty actually is, we see that novelty happens simply because it can. The idea of "cosmic necessity" becomes completely meaningless in the case of any specific particular novelty we might care to examine.

The question "Why does novelty exist?" is not coherent enough to be meaningfully answerable. But we still feel the intuitive temptation to ask the question, even if it involves stepping outside the bounds of meaningful language. The emotional temptation human beings seem to have to wrap up the answer to this particular why-question in some sort some of "creator" with socio-political or ethical relevance borders on laughable. There is a clear disconnect between the historical narratives that have been constructed by human religion and the compelling nature of the evidence for intermittent novelty that has been provided by 21st century historical science. But where does this leave our inclination to ask "Why does novelty exist?"? As something to be met with intellectual laziness and blind faith, to be answered with something along the lines of "Only God knows"? As a relic of our superstitious past, to be sneered at, suppressed, and forgotten about? Answering the question "Why does novelty exist?" by either "Novelty exists by virtue of an underlying creative intelligence in the universe." or "The 'question' as to the reason why novelty exists is, in fact, a confused non-question, due to the anthropic principle." each, in different ways, amounts to a meaningless cop-out, a lumping of the uncertainties within the question into poorly defined words.

Novelty might perhaps be one of the things that we would be best advised to pass over in reflective, perhaps reverent, silence. But perhaps not. The natural, novelty-driven temporal evolution of meaningful language, suggests that we should not be so puritanical about dismissing questions that are, strictly speaking, unanswerable. The boundary separating the ineffable from the meaningful is not fixed. There is potentially more to say, even if we are not yet ready to say it. The question does not go away or become insignificant by virtue of its being unanswerable. Answerability and meaningfulness flow with time - due to the newness that is manifest in existence. Why does novelty exist?

"*What the imagination seizes as beauty must be truth - whether it existed before or not.*"
John Keats, 1817

BIBLIOGRAPHY

Books

Peter Atkins. *The laws of thermodynamics: A very short introduction.* Oxford University Press. 2010.
Alain Badiou. *Being and Event.* (Translated by Oliver Feltham). Bloomsbury Academic. 2013 (First published 1988).
Michael Behe. *Darwin's black box: The biochemical challenge to evolution.* Free Press. 2006.
Michael Behe. *The edge of evolution: The search for the limits of Darwinism.* Free press. 2007.
Jeremy Berg, John Tymoczko, Gregory Gatto & Lubert Stryer. *Biochemistry: 8th edition.* Freeman. 2015.
Henri Bergson. *Creative Evolution.* Forgotten Books. 2015. (First published 1907).
Henri Bergson. *Time and Free Will: An essay on the immediate data of consciousness.* Dover. 2001 (First published1910).
Henri Bergson. *Matter and Memory.* HardPress.2012. (First published1912).
Simon Blackburn. *The Oxford dictionary of Philosophy.* Oxford University Press. 2008.
William Blattner. *Heidegger's Being and Time.* Continuum. 2006.
David Bohm. *Wholeness and the Implicate Order.* Routledge. 1980.
Lee Braver. "*Groundless grounds: A study of Wittgenstein & Heidegger.* MIT press. 2012.
Donald Canfield. *Oxygen: A four billion year history.* Princeton University Press. 2014.
Stanley Cavell. *The claim of reason.* Oxford University Press. 1979.
Arthur C. Clark. 2001*: A space odyssey.* Hutchinson. 1968.
Simon Conway Morris. *Life's solution: Inevitable humans in a lonely universe.* Cambridge University Press. 2003.
Alice Crary & Rupert Read. *The new Wittgenstein.* Routledge. 2000.
Francis Crick. *Life itself: It's origin and nature.* Futura Publications. 1982.
Donald Crosby. *Novelty.* Lexington.2005.
Charles Darwin. *On the origin of species by means of natural selection: or the Preservation of the Favoured Races in the Struggle for Life.* John Murray. 1857.
Richard Dawkins. *The extended phenotype.* Oxford University Press. 1982.
Christian de Duve. *Vital Dust: The origin and evolution of life on Earth.* Basic Books. 1995.
Christian de Duve. *Life Evolving: Molecules, mind and meaning.* Oxford University Press. 2002.
William Dembski. *The design inference: Eliminating chance through small probabilities.* Cambridge University Press. 1998.
William Dembski. *Intelligent Design: The Bridge between science and theology.* Intervarsity Press. 1999.
William Dembski. *No Free Lunch: Why specified complexity cannot be purchased without intelligence.* Rowman & Littlefield. 2001.
Rene Descartes. *Specimina philosophiae.* Elsevier. 1644.
Hubert Dreyfus. *What Computers Can't Do: The limits of artificial intelligence.* HarperCollins. 1978.

Hubert Dreyfus, H. *What computers still can't do: A critique of artificial reason.* MIT press. 1992.
David Edmonds & John Eidinow. *Wittgenstein's poker: The story of a ten minute argument between two great philosophers.* Harper Perrenial, 2002.
Arthur Fairbanks. *Empedocles: Fragments and Commentary.* Create Space. 2012.
Paul Feyerabend. *Farewell to reason.* Verso Books. 1987.
Ronald Fisher. *The Genetical Theory of Natural Selection.* Oxford University Press. 1930
Hans-Geog Gadamer. *The beginning of Philosophy.* Bloomsbury Academic. 2000.
Anthony Grayling. *Wittgenstein: A very short introduction.* Oxford University Press. 1996.
Mark Haddon. *The curious incident of the dog in the night time.* Vintage. 2004.
Martin Heidegger. *Being and Time.* (Translated by J. Macquarrie & E. Robinson). Blackwell 1962 (First published 1927).
David Hume. *An enquiry concerning Human Understanding.* Hackett Classics. 1993 (First published 1748).
Michael Inwood. *Heidegger: A very short introduction.* Oxford University Press. 2000.
Eva Jablonka & Marion Lamb. *Evolution in four dimensions: Genetic, epigenetic, behavioural and symbolic variation in the history of life.* Bradford Books. 2014.
Carl Jung. *Synchronicity: A acausal connecting principle.* Princeton University Press. 1960.
Immanuel Kant. *Critique of pure reason.* Cambridge University Press. 1999. (First published 1781).
Immanuel Kant. *Critique of judgement.* (Translated by Werner Pluhar). Hackett Classics. 1987. (First published 1790).
James Kasting. *How to find a habitable planet.* Princeton University Press. 2010.
Stuart Kauffman. *The Origins of Order: Self Organization and Selection in Evolution.* Oxford University Press. 1993.
Boris Kozo-Polyanski. *Symbiogenesis: A new principle of evolution.* (Translated by Lynn Margulis & Victor Fet). Harvard University Press. 2010 (First published 1924).
Andrew Knoll. *Life on a young planet: The first three billion years of evolution on Earth.* Princeton University Press. 2003.
Helge Kragh. *Cosmology and Controversy: The historical development of two theories of the Universe.* Princeton University Press. 1999.
Gerhardt B. Ladner. *The idea of reform: Its impact on Christian thought and action in the age of the fathers.* Literary Licensing 2011.
Nick Lane. *Mitochondria and the meaning of life.* Oxford University Press. 2005.
F. William Lawvere & Stephen Schanuel. *Conceptual Mathematics: A first introduction to Categories.* Cambridge University Press 2009.
Timothy Lenton & Andrew Watson. *Revolutions that made the Earth.* Oxford University Press. 2011.
George Lewes. *Problems of Life and Mind.* London: Trübner. 1875.
James Lovelock. *Gaia: A new look at life on Earth.* Oxford University Press. 1979.
James Lovelock. *The Ages of Gaia: A biography of our living Earth.* Norton. 1988.
James Lovelock. *The Vanishing Face of Gaia.* Basic Books. 2009.
Lynn Margulis. *Symbiosis in cell evolution.* Freeman. 1981

Lynn Margulis. & Dorion Sagan. *Microcosmos: Four billion years of microbial evolution*. University of California Press. 1986.
Lynn Margulis. & Dorion Sagan. *What is life?* University of California Press. 1995.
Lynn Margulis. *Symbiotic Planet: A new look at evolution*." Basic books. 1999.
John Maynard-Smith & Eors Szathmary. *The major transitions in evolution*. Oxford University Press. 1995.
Terrence McKenna & Dennis McKenna. *The invisible landscape: Mind, Hallucinogens and the I Ching*. HarperOne. 1994.
Denis McManus. *The enchantment of words: Wittgenstein's Tractatus Logico Philosophicus*. Oxford University Press. 2006.
Daniel McShea & Robert Brandon. *Biology's First Law: The tendency for diversity and complexity to increase in evolutionary systems*. University of Chicago Press. 2010.
Gregor Mendel. *Experiments in plant hybridization*. Cosimo Classics. 2008. (First published 1865).
Richard Michod. *Darwinian Dynamics: Evolutionary transitions in fitness and individuality*. Princeton University Press. 2000.
Ray Monk. *Ludwig Wittgenstein: The duty of genius*. Penguin Books. 1991.
C. Lloyd Morgan. *Emergent Evolution*. Henry Holt. 1923.
Harold Morowitz. *Beginnings of cellular life: Metabolism recapitulates biogenesis*. Yale University Press. 1984.
Harold Morowitz: *The emergence of everything: How the world became complex*. Oxford University Press. 2004.
Alexander P.D. Mourelatos (Ed). *The Pre-Socratics: A collection of critical essays*. Princeton University Press. 2014.
Thomas Nagel. *Mind and Cosmos: Why the Materialist Neo-Darwinian Conception of Nature is Almost Certainly False*. Oxford University Press. 2012.
Friedrich Nietzsche. *Thus spoke Zarathustra: A book for all and for none*. (Translated by R.J. Hollingdale). Penguin Classics 1974.(First published 1883).
Friedrich Nietzche. *The will to power*. (Translated by W. Kauffmann & R.J. Hollingdale). Vintage Books 1967. (First published1901).
Michael North. *Novelty: A History of the New*. University of Chicago Press. 2013.
Hayden Pellicia (Ed), & Bejamin Jowett (trans). *Plato: Complete works. The Benjamin Jowett translation*. Modern Library Classics. 2001.
Roger Penrose. *Cycles of time: An extraordinary new view of the universe*. Vintage (reprint edition). 2012.
Heinrich Wiegand Petzet. *Encounters and Dialogues with Martin Heidegger, 1929-1976*. University of Chicago Press.1993.
Mark Ridley (Ed). *Evolution* (3rd Edition). Wiley-Blackwell. 2003.
Bertrand Russell *The Principles of Mathematics*. Norton. 1938. (First published 1903).
Bertrand Russell. *The Philosophy of Bergson*. Cambridge, London : Macmillan and Co. Ltd. 1914.
Bertrand Russell. *Introduction to Mathematical Philosophy*. Routledge. 1993 (First published 1919).
Bertrand Russell. *Skeptical Essays*. Routledge. 1928.
Bertrand Russell *Wisdom of the West*. Crescent Books. 1989.
Joe Sachs. *Aristotle's Metaphysics*. Trade Select. 1999.
Paul Schlipp. *The Philosophy of Rudolf Carnap*. Open Court Publishing Company 1999.

Erwin Schrondinger. *What is life? The physical aspect of the living cell.* Cambridge University Press. 1944.
Dirk Schulze-Makuch & Louis Irwin. *Life in the universe: Expectations and constraints.* Springer. 2006.
David Schwartzmann. *Life, temperature and the Earth.* Columbia University Press. 1999.
Rupert Sheldrake. *A new science of life: The hypothesis of morphic resonance.* Park Street Press.1995.
Rupert Sheldrake. *The presence of the past: Morphic resonance and the habits of nature.* Park Street Press. 2012.
Rupert Sheldrake. *The Science delusion.* Coronet. 2012.
Raymond Smullyan & Melvin Fitting. *Set Theory and the Continuum Problem.* Dover. 2010.
Eilliot Sober & David Sloan-Wilson. *Unto others: The evolution and psychology of unselfish behaviour.* Harvard University Press. 1998.
Kim Sterelny & Paul Griffiths. *Sex and Death: An Introduction to the Philosophy of Biology.* University of Chicago Press. 1999.
Paul Strathern. *Heidegger in 90 minutes.* Ivan R. Dee. 2002.
Michael Thain & Michael Hickman. *The Penguin dictionary of biology.* Penguin Books. 2004.
Toby Tyrrell. *On Gaia: A critical investigation into the relationship between life and Earth.* Princeton University Press. 2013.
Vladimir Vernadsky. *The Biosphere.* Synergetic Press. 1986. (First published 1926).
Tyler Volk. *Gaia's Body: Towards a physiology of Earth.* MIT Press. 2003.
Conrad Waddington. *The ethical animal.* George Allen & Unwin. 1960.
Conrad Waddington. *The nature of life.* George, Allen & Unwin. 1961.
Peter Ward. *The Medea Hypothesis: Is life on Earth ultimately self-destructive?* Princeton University Press. 2015.
Robin Waterfield. *The first philosophers: The pre-socratics and sophists.* Oxford World's Classics. 2009.
Weismann, A. *The germ plasm: A theory of heredity.* Charles Schriber's sons. (Electronic scholarly publishing http://www.esp.org/books/weismann/germ-plasm/facsimile/). 1893.
Alfred North Whitehead. *Process and Reality: An essay in cosmology.* MacMillan. 1929.
Robert Williams & Joao da Silva. *The chemistry of evolution: The development of our ecosystem.* Elsevier. 2006.
Hermeine Wittgenstein & Fania Pascal. *Recollections of Wittgenstein.* Oxford University Press. 1984.
Ludwig Wittgenstein. *Tractatus Logico Philosophicus.* Annalen der Naturphilosophie (W. Ostwald). 1918 (1921).
Ludwig Wittgenstein. *Philosophical investigations.* Prentice Hall. 1953.
Ludwig Wittgenstein. *Remarks on the Foundations of Mathematics.* Wiley Blackwell. 1981(First published 1953).
Ludwig Wittgenstein. *Zettel.* Wiley Blackwell. 1981(First published 1968)
Ludwig Wittgenstein. *On certainty.* Harper and Row. 1953.
Bernard Wood. *Human Evolution: A very short introduction.* Oxford University Press. 2006.

Journal Articles

Amelin, Y., Gosh, A. & Rotenburg, E. *Unraveling the evolution of chondrite parent asteroids by precise U-Pb dating and thermal modeling.* Geochimica Cosmochimica Acta. 69(2). 505 – 518. 2005.

Allison, A.C. *Genetic control of resistance to human malaria.* Current Opinion in Immunology 21 (5): 499–505. 2009.
Bernays, P. *Comments on Ludwig Wittgenstein's Remarks on the Foundations of Mathematics.* Ratio, 2 (1): 1–22. 1959.
Boyle, R.A., Lenton, T.M. & Williams H.T.P. *Neoproterozoic "snowball Earth" glaciations and the evolution of altruism.* Geobiology 5(4). 337-349. 2007.
Boyle, R.A., Lenton, T.M. & Watson, A.J. *Symbiotic physiology promotes homeostasis in daisyworld.* Journal of theoretical biology 274.170 – 182. 2011.
Boyle, R.A., Williams, H.T.P & Lenton, T.M. *Natural selection for costly nutrient recycling in simulated microbial metacommunities.* Journal of theoretical biology 312. 1 – 12. 2012.
Boyle, R.A. *The problem of eukaryotic origins in relation to the early/mid Proterozoic Earth System.* In *Revolutions in the Early Proterozoic: Tracking Geochemical and Geobiological Change* (Edited by S. Poulton & D. Johnston D). Springer. In Press.
Cantor, G *Ueber eine elementare Frage der Mannigfaltigkeitslehre.* Jahresbericht der Deutschen Mathematiker-Vereinigung 1890–1891.
Carter, B. *Large Number Coincidences and the Anthropic Principle in Cosmology.* IAU Symposium 63: Confrontation of Cosmological Theories with Observational Data. Dordrecht: Reidel. pp. 291–298. 1974.
Cech, T. 2000. *The ribosome is a riobozyme.* Science 289.878 – 879. 2000.
Chomsky, N. *Three Models for the Description of Language.* IRE Transactions on Information Theory 2 (3): 113–124. 1956.
Connelly, J.N., Bizzarro, M, Krot, A.N., Nordlund, A, Wielandt, D. & Ivanova, M.A. *The Absolute Chronology and Thermal Processing of Solids in the Solar Protoplanetary Disk.* Science 338. 651-655. 2012.
Cook, N.D. *The case for reverse translation.* Journal of theoretical biology 64(1). 113 – 135. 1977.
Crick F.H. *The origin of the genetic code.* J Mol Biol 38 (3): 367–379. 1968.
Dale, T. *Protein and nucleic acid together: A mechanism for the emergence of biological selection.* Journal of Theoretical Biology 240.337 – 342. 2006.
Dill, K.A. & McCallum, J.L. *The Protein folding problem: 50 years on.* Science 338(6110) 1042 – 1046.2012.
Doolittle, W. F. *Is Nature Really Motherly?* Coevolution Quarterly 29: 58–63. 1981.
Eigen, M. *Self organisation of matter and the evolution of biological macromolecules.* Naturwissenschaften 64.541 – 565.1971.
Fisher, R.A. *The correlation between relatives on the supposition of Mendelian inheritance.* Trans. Royal. Soc. Edinburgh 52.399 – 433. 1918.
Frank, S.A. *Models of symbiosis.* American Naturalist, 150.S1. *S*80 – *s*99. 1997.
Ganti, T. *Organisation of chemical reactions into dividing and metabolising units: the chemotons.* Acta Chimica Academia Scientarium Hungarica98.265 – 283. 1975.
Gianoli, E, & Carrasco-Ura, F. *Leaf mimicry in a climbing plant protects against Herbivory.* Current Biology 24(9). 984 – 987. 2014.
Gilbert, W. *The RNA World.* Nature 319 (6055): 618. 1986.
Godel, K. *On Formally Undecidable Propositions of Principia Mathematica and Related Systems.* Monatshefte für Mathematik. 1931.
Goodnight, C.J. *Multi-level selection: The evolution of cooperation in non-kin groups.* Population Ecology 47:3–12. 2005.

Graeber, D. *Of flying cars and the declining rate of profit.* The Baffler, 19. 2012.
Griffths, P. *Genetic information: A metaphor in search of a theory.* Philosophy of Science 68. 394 – 412. 2001.
Hamilton, W. D. *The evolution of altruistic behavior.* American Naturalist 97 (896): 354–356. 1963.
Harvey, I. *Homeostasis and rein control: from Daisyworld to active perception.*" Ninth International Conference on the Simulation and Synthesis of Living Systems. MIT Press, pp.309–314.2004.
Hintikka, J. *Wittgenstein on Being and Time.* Ludwig Wittgenstein: Half truths and one and a half truths. 241 – 247. 1996.
Hintikka, J. *What does the Wittgensteinian inexpressible express?* The Harvard Review of Philosophy *XI*. 2003.
Hofmann, H.J.; Narbonne, G.M.; Aitken, J.D. *Ediacaran remains from intertillite beds in northwestern Canada.* Geology 18 (12): 1199–1202. 1990.
Hoffman, P.F. & Schragg., D.P. *Snowball Earth: Testing the limits of global change.* Terra Nova. 14(3). 129 – 155. 2002.
Javaux, E.2011 *Origins and Evolution of Life: An Astrobiological Perspective.* 414–449.2011.
Kirschvink, J.L. *Late Proterozoic Low-Latitude Global Glaciation: the Snowball Earth.* The Proterozoic Biosphere: A Multidisciplinary Study. J.W. Schopf & C.Klein. Cambridge University Press. pp51 – 52 . 1992.
Lenton, T.M., Boyle, R.A., Poulton, S.W., Shields-Zhou, G.A. & Butterfield, N.J. *Co-evolution of eukaryotes and ocean oxygenation in the Neoproterozoic era.* Nature Geoscience 7 (4), 257-265. 2014.
Lovelock, J.E. *Gaia as seen through the atmosphere.* Atmospheric Environment. 6. 579 – 580. 1972.
Lovelock, J.E. & Margulis, L.M. *Atmospheric homeostasis by and for the biosphere: the gaia hypothesis.* Tellus XXVI 1 – 2. 2 – 10. 1973.
Lovelock, J.E. *Hands up for the Gaia hypothesis.* Nature 3(44). 100 – 102. 1990.
McLaughlin, P. *Kant's critique of Teleology in Biological explanation: Antinomy and Teleology.* Studies in the History of Philosophy 16, 1990.
Miller, S.L.& Urey, H.C. *Organic Compound Synthesis on the Primitive Earth.* Science 130 (3370): 245–51. 1959.
Meyer, S.C. *Not by chance: From bacteria propulsion systems to human DNA, evidence of intelligent design is everywhere.* National Post of Canada, December 10th 2005.
Mira, A. Martin-Cuadrado, A.B., D'Auria, G. & Rodriguez-Valera, F. *The bacterial pan-genome: A new paradigm in microbiology.* Int. Microbiol. 13. 45 – 57. 2010.
Moore, G.E. *A defence of common sense.* Contemporary British Philosophy (2nd Series) 1925
Nashimoto, M. *The RNA/Protein Symmetry hypothesis: Experimental support for reverse translation of primitve proteins.* Journal of theoretical biology 290(2). 181 – 187. 2001.
Nirenberg, R.L. & Nirenberg, D. *Badiou's Number: A Critique of Mathematics as Ontology.* Critical Inquiry 37, University of Chicago. 2011.
Nowak, M.A. *Five rules for the evolution of cooperation.* Science. 314(5805) 1560-1563. 2006.
Orgel L.E. *Evolution of the genetic apparatus.* J Mol Biol 38 (3): 381–393. 1968.
Pigliucci, M. *Chance, necessity, and the war against science.* BioScience 50: 79–72. 2000.
Prusiner, S.B. *Prions.* Proceedings of the National Academy of Sciences of the United States of America 95 (23): 13363. 1998.

Robertson M.P., Joyce G.F. *The origins of the RNA world.* Cold Spring Harb Perspect Biol 4 (5). 2012.
Russell, B. *Mathematical logic as based on the theory of types.* American Journal of Mathematics, 30: 222–262. 1908.
Saunders, P.T., Koeslag, J.H. & Wessels,J.A. *Integral rein control in physiology.* J. Theor.Biol.194,163–173. 1998.
Scruton, R. *A Nothing Would do As Well.* Times Literary Supplement. 31st August 2012.
Searle, J. *The Word Turned Upside Down*, review of *On Deconstruction: Theory and Criticism after Structuralism*, The New York Review of Books. 1983.
Shannon, C. E. *A Mathematical Theory of Communication.* Bell Systems Technical Journal 27 (3): 379–423. 1948.
Spiegelman, S. *Extracellular evolution of replicating molecules.* In *The Neuro Sciences: A Second study program*, pp 927 – 945 Schmidtt, F.O. ed. Rockefeller University Press. 1970.
Stace, W.T. *Novelty, indeterminism and emergence.* The philosophical review 48. 296 – 310. 1939.
Van Baalen, M. & Rand, D.A. *The unit of selection in viscous populations and the evolution of altruism.* Journal of theoretical biology 193(4):631-648. 1998.
Waddington, C.H. *Canalization of development and the inheritance of acquired characters.* Nature, 150, 563-565. 1942.
Waggoner, B. *Interpreting the Earliest Metazoan Fossils: What Can We Learn?.* Integrative and Comparative Biology 38 (6): 975–982. 1998.
Watson J.D. & Crick F.H. *A Structure for Deoxyribose Nucleic Acid.* Nature 171 (4356): 737–738. 1953.
Watson, A.J.; J.E. Lovelock. *Biological homeostasis of the global environment: the parable of Daisyworld.* Tellus B (International Meteorological Institute) 35 (4): 286–9. 1983.
Wilson, D.S. & Sober, E. *Reintroducing group selection to the human behavioral sciences.* Behavioral and Brain Sciences 17 (4): 585–654. 1994.
Wolpert, L & Sheldrake, R. *What can DNA tell us? Place your bets now.* New Scientist, 8th July 2009.

Additional notes

Opening quotations:

1. The Rigveda, (Ancient Indian collection of Verdic Sanskrit hymns, probably composed during the approximate interval 1100 – 1700B.C.). https://en.wikipedia.org/wiki/Rigveda
2. Parmenides. *Commentary on the physics.* Frag. B 8.1 – 4 (~5th century B.C.).
3. *Holy Bible.* English Standard version. (Old Testament composed approximated 3rd century B.C.)
4. Ludwig Wittgenstein. *Tractatus Logico Philosophicus.* Annalen der Naturphilosophie (W. Ostwald). 1918 (1921).
5. Heidegger, M. *What is Metaphysics?* Inaugural lecture at University of Freiburg.1929

Closing quotation:

1. Keats, J. Letter to Benjamin Bailey (November 22, 1817).

www.ingramcontent.com/pod-product-compliance
Ingram Content Group UK Ltd.
Pitfield, Milton Keynes, MK11 3LW, UK
UKHW050945290926
15470UKWH00005B/1650